T0229533

Advances in Soft Computing
Applications

RIVER PUBLISHERS SERIES IN MATHEMATICAL, STATISTICAL AND COMPUTATIONAL MODELLING FOR ENGINEERING

Series Editors:

MANGEY RAM
Graphic Era Deemed to be University, India

TADASHI DOHI
Hiroshima University, Japan

ALIAKBAR MONTAZER HAGHIGHI
Prairie View Texas A& M University, USA

Applied mathematical techniques along with statistical and computational data analysis has become vital skills across the physical sciences. The purpose of this book series is to present novel applications of numerical and computational modelling and data analysis across the applied sciences. We encourage applied mathematicians, statisticians, data scientists and computing engineers working in a comprehensive range of research fields to showcase different techniques and skills, such as differential equations, finite element method, algorithms, discrete mathematics, numerical simulation, machine learning, probability and statistics, fuzzy theory, etc.

Books published in the series include professional research monographs, edited volumes, conference proceedings, handbooks and textbooks, which provide new insights for researchers, specialists in industry, and graduate students. Topics included in this series are as follows:-

- Discrete mathematics and computation
- Fault diagnosis and fault tolerance
- Finite element method (FEM) modeling/simulation
- Fuzzy and possibility theory
- Fuzzy logic and neuro-fuzzy systems for relevant engineering applications
- Game Theory
- Mathematical concepts and applications
- Modelling in engineering applications
- Numerical simulations
- Optimization and algorithms
- Queueing systems
- Resilience
- Stochastic modelling and statistical inference
- Stochastic Processes
- Structural Mechanics
- Theoretical and applied mechanics

For a list of other books in this series, visit www.riverpublishers.com

Advances in Soft Computing Applications

Editors

Shristi Kharola

Graphic Era Deemed to be University, India

Mangey Ram

Graphic Era Deemed to be University, India
Peter the Great St. Petersburg Polytechnic University, Russia

Sachin K. Mangla

O. P. Jindal Global University, India
University of Plymouth, United Kingdom

Yigit Kazancoglu

Yasar University, Turkey

NEW YORK AND LONDON

Published 2023 by River Publishers
River Publishers
Alsbjergvej 10, 9260 Gistrup, Denmark
www.riverpublishers.com

Distributed exclusively by Routledge
605 Third Avenue, New York, NY 10017, USA
4 Park Square, Milton Park, Abingdon, Oxon OX14 4RN

Advances in Soft Computing Applications / by Shristi Kharola, Mangey Ram, Sachin K. Mangla, Yigit Kazancoglu.

Routledge is an imprint of the Taylor & Francis Group, an informa business

ISBN 978-87-7022-817-6 (print)
ISBN 978-87-7022-961-6 (paperback)
ISBN 978-10-0092-206-6 (online)
ISBN 978-1-003-42588-5 (ebook master)

While every effort is made to provide dependable information, the publisher, authors, and editors cannot be held responsible for any errors or omissions.

Contents

Preface

Until recently, machines could only recognize numbers, alphabets, words, and symbols. They were trained to think by making decisions like yes or no, true or untrue, high or low, and so on. However, while discussing a machine that can retain information like a human mind and exhibit intelligent conduct like people, it must also evolve relevantly in order to properly function like humans. The propensity of today's technologies to think like humans may be witnessed in new developing areas such as neural computing, fuzzy logic, evolutionary computation, machine learning, and probabilistic reasoning. These collections of techniques coincide in one principal technique known as soft computing. Soft computing seeks to formalize our cognitive processes by using the human mind as a model. It is a subfield of artificial intelligence that explores the approaches used to solve problems using imprecise, unpredictable, and/or inaccurate data.

"Uncertainty" is unquestionably a major feature of human thinking. This idea of how to convey "uncertainty" in programming resulted in the development of a theory known as fuzzy theory. The concept of fuzziness focuses around the representation of ideas that are somewhat ill-defined, unclear, or, as the term implies, uncertain. As a way, fuzzy theory may be said to be represented in such a manner that it describes both randomness and uncertainty at the same time. Fuzzy sets and soft computing offer multiple theoretical and practical tools for challenging linguistic and numerical modeling applications. When human judgments and modeling of human knowledge are required, fuzzy set techniques are typically the best choice.

This book offers the most recent soft computing and fuzzy-logic-based applications and developments in the areas of industrial advancements, supply chain and logistics, system enhancement, decision-making, artificial intelligence, smart systems, and other rapidly evolving technologies. The book introduces soft computing applications in today's competitive environment to assist industries in overcoming the challenges brought by sophisticated judgment making systems. The mathematical foundation as well as the most important applications of this theory to other theories and methodologies will indeed be explained in this book. However, in this review, we shall concentrate largely on fuzzy set theory. Examining both research and practical dimensions, the book includes contributions authored by current researchers and/or experienced practitioners in the area, bridging the gap between theory and practice and addressing new research problems in systems enhanced with soft computing techniques. All the chapters in the book are authored by leading scholars and practitioners in their respective fields of expertise and provide a diversity of novel methodologies, innovations, and solutions that have not previously been discussed in the literature.

Editors:

Shristi Kharola
Graphic Era Deemed to be University, India

Mangey Ram
Graphic Era Deemed to be University, India
Peter the Great St. Petersburg Polytechnic University, Saint Petersburg, Russia

Sachin K. Mangla
O. P. Jindal Global University, Sonipat, India
University of Plymouth, United Kingdom

Yigit Kazancoglu
Yasar University, Turkey

Acknowledgement

This book is based on the research conducted in the excelling areas of soft computing and fuzzy logic. It aims to unify existing works and, to this end, several research directions have been investigated. The editors are thankful to the River Publishers for their time and management toward the book. They would also like to acknowledge authors and reviewers who assisted in the production of this book. This book is very theoretical and mathematical. It may be useful to research workers wishing to gain a broad view of the various developments and those researching into particular applications of soft computing. The book is very well printed: a tribute to the publishers bearing in mind the high incidence of Greek and mathematical symbols. Thank you once again to everyone who strives to grow and help others grow.

Editors

List of Figures

List of Tables

List of Contributors

Afonso, Shona, *Department of Computer Engineering, Padre Conceicao College of Engineering, India*

Agarwal, Mohini, *Amity School of Business, Amity University, India*

Anand, Adarsh, *Department of Operational Research, University of Delhi, India*

Arora, H. D., *Department of Mathematics, Amity Institute of Applied Sciences, Amity University, India*

Avikal, Shwetank, *Department of Management Studies, Graphic Era Hill University, India*

Bairagi, Bipradas, *Department of Mechanical Engineering, Haldia Institute of Technology, India*

Bhagavatham, Hari Krishna, *OUCCBM, Osmania University, India*

Bhatia, Mansi, *Department of Mathematics, Amity Institute of Applied Sciences, Amity University, India*

Bocharov, Mykhailo, *Research Laboratory Moral and Psychological Support of Troops (Forces) Activity Department, National Defense University of Ukraine named after Ivan Cherniakhovskyi, Ukraine*

Bose, Goutam, *Department of Mechanical Engineering, Haldia Institute of Technology, India*

Chachra, Aayushi, *Graphic Era Deemed to be University, India*

Chakravaram, Venkamaraju, *Jindal Global Business School, O.P. Jindal Global University, India*

Chanchal, *Department of Operational Research, University of Delhi, India*

Dey, Balaram, *Department of Mechanical Engineering, Haldia Institute of Technology, India*

Dumitrescu, C., *Splaiul Independentei, Bucharest, Romania*

Filipishyna, Liliya, *National University of Shipbuilding named after Admiral Makarov, Ukraine*

Goyal, Nupur, *Graphic Era Deemed to be University, India*

Kaminsky, Oleg, *Department of Computer Mathematics and Information Security, Kyiv National Economic University named after Vadym Hetman, Ukraine*

Kazancoglu, Yigit, *Department of Logistics Management, Yasar University, Turkey*

Kharola, Shristi, *Graphic Era Deemed to be University, India*

Kotlubai, Viacheslav, *National University "Odessa Law Academy," Ukraine*

Koval, Viktor, *National Academy of Sciences of Ukraine, Ukraine*

Kumar, Akshay, *Graphic Era Hill University, India*

Kumar, Deepak, *Department of Mathematics, D. S. B. Campus, Kumaun University, India*

Kumar, Hitesh, *Department of Operational Research, University of Delhi, India*

Kumar, Pawan, *Department of Mathematics, Government Degree College, India*

Kumar, Vimal, *Department of Information Management, Chaoyang University of Technology, Taiwan*

Malik, Arpit, *Jindal Global Business School, O.P. Jindal Global University, India*

Malik, P., *Graphic Era (Deemed to be) University, India*

Mittal, V., *Graphic Era (Deemed to be) University, India*

Naithani, Anjali, *Department of Mathematics, Amity Institute of Applied Sciences, Amity University, India*

Nithin Kumar, K. C., *Department of Mechanical Engineering, Graphic Era Deemed to be University, India*

Pai, Anusha, *Department of Computer Engineering, Padre Conceicao College of Engineering, India*

Pant, Rushali, *Department of Mechanical Engineering, Graphic Era Hill University, India*

Purohit, K. C., *Graphic Era (Deemed to be) University, India*

Ram, Mangey, *Graphic Era Deemed to be University, India; Institute of Advanced Manufacturing Technologies, Peter the Great St. Petersburg Polytechnic University, Russia*

Ratnakaram, Sunitha, *Jindal Global Business School, O.P. Jindal Global University, India*

Tsimoshynska, Oksana, *Interregional Academy of Personnel Management, Ukraine*

Vdovenko, Nataliia, *National University of Life and Environmental Sciences of Ukraine, Ukraine*

Verma, Raksha, *Shaheed Rajguru College of Applied Sciences for Women, University of Delhi, India*

Verma, Riya, *Miranda House, University of Delhi, India*

Yankovyi, Oleksandr, *Odessa National Economic University, Ukraine*

Yereshko, Julia, *Department of Economic Cybernetics, National Technical University of Ukraine "Igor Sikorsky Kyiv Polytechnic Institute," Ukraine*

List of Abbreviations

AHP	Analytical hierarchy process
AI	Artificial intelligence
ANFIS	Adaptive neuro-fuzzy inference system
BIM	Building information model
BNP	Best non-fuzzy performance
BRF	Base rule fuzzy
BRF	Fuzzy rule base
BSC	Balanced scorecard
BW-EDAS	Best-worst and evaluation based on distance from average solution
CBR	Case-based reasoning
CDSS	Clinical decision support system
CoA	Center of area
CoM	Center of maximum
COMET	Characteristic objects method
DoA	Direction-of-Arrival
DSS	Decision support system
EEG	Electroencephalogram
EROI	Energy return on investment
FACCT	Fuzzy applied cell control technology
FAHP	Fuzzy analytical hierarchical procedure
FAHP	Fuzzy analytical hierarchy process
FAM	Fuzzy associative memory
FAT	Fuzzy approximation theorem
FDT	Fuzzy decision tree
FES	Fuzzy expert system
FFBAT	Firefly-BAT
FGT	Fuzzy gray theory
FIS	Fuzzy inference system
FL	Fuzzy logic
FLC	Fuzzy logic controller
FLIPS	Fuzzy flip-flop

FMCDM	Fuzzy multiple criteria decision making
FMOORA	Fuzzy multi-objective optimization by ratio analysis
FNIS	Fuzzy negative ideal solution
FPGA	Field-programmable gate array
FPIS	Fuzzy positive ideal solution
FTOPSIS	Fuzzy- Technique for order of preference by similarity to ideal solution
FVIKOR	Fuzzy- VlseKriterijumska Optimizacija I Kompromisno Resenje
GA	Genetic algorithm
GDM	Group decision-making
GRC	Gray relational coefficient
GRG	Gray relational grade
IFS	Intuitionistic fuzzy set
IoT	Internet of Things
KNN	K-nearest neighbor
LCSA	Life cycle sustainability assessment
LIFE	Laboratory for International Fuzzy Engineering
MCDA	Multi-criteria decision analysis
MCDM	Multi-criteria decision making
MCI	Mild cognitive impairment
MH	Malignant hyperpyrexia
MoM	Mean of maximum
MOORA	Multi-objective optimization by ratio analysis
MOPA	Multi-objective performance analysis
NLP	Natural language processing
NM	Non-membership
NRES	Non-renewable energy systems
PD	Parkinson's disease
PF	Programming flexibility
PFS	Pythagorean fuzzy set
PI	Performance index
PID	Proportional-integral-derivative
PLC	Programmable logic controller
PROMETHEE II	Preference ranking for organization method for enrichment evaluation-II
ReLU	Rectified linear unit
RES	Renewable energy systems

RF	Random forest
ROI	Return on investment
RSA	Rivest-Shamir-Adleman
SAW	Simple additive weighting
SCM	Supply chain management
SRAM	Static random-access memory
TFN	Triangular fuzzy number
TFR	Fuzzy-real transformation
TOPSIS	Technique for order of preference by similarity to ideal solution
TRF	Transformation real-fuzzy
VIKOR	VIseKriterijumska Optimizacija I Kompromisno Resenje
VSQ	Vendor service quality
WEF	World Economic Forum

1

Applying Fuzzy Logic to the Assessment of Latent Economic Features

Oleksandr Yankovyi[1], Oksana Tsimoshynska[2], Viktor Koval[3], Yigit Kazancoglu[4], Viacheslav Kotlubai[5], and Liliya Filipishyna[6]

[1]Odessa National Economic University, Ukraine
[2]Interregional Academy of Personnel Management, Ukraine
[3]National Academy of Sciences of Ukraine, Ukraine
[4]International Logistics Management, Yasar University, Turkey
[5]National University "Odessa Law Academy," Ukraine
[6]National University of Shipbuilding named after Admiral Makarov, Ukraine

Abstract

The impossibility of direct measurement of some economic features in the metric scale prompted researchers to use specific mathematical and statistical apparatus in their evaluation, in particular the combination of expert scores with models and methods of fuzzy logic. Such latent features at the micro-level include financial condition, competitiveness, investment attractiveness, the global criterion of optimality, priority of the company's investment projects, etc. The concepts and essence of latent economic signs are discussed, their definition and connection with scales of measurement of indicators of commodity producers are given, and also the four-level hierarchical model of latent sign "competitiveness" of the company is investigated. The methods of estimation of latent economic signs existing in modern science are considered and their classification is offered. In the example of the sign of "competitiveness" of the company, methods of estimating latent indicators based on a combination of expert opinions and algorithms of fuzzy sets and fuzzy logic are considered. The method of determining the weights of the symptom factors of the desired latent economic indicator based on their pairwise comparison is discussed (Saaty process).

Keywords: expert evaluation, Fuzzy set, fuzzy logic, latent feature, linguistic variable, membership function.

1.1 Introduction

It is traditionally believed that the use of mathematics and statistics in economics is manifested in obtaining only quantitative characteristics of the studied objects. This understanding seems rather simplistic because quantitative definitions are always associated with qualitative properties. Specific research is conducted at different levels of economics: at the level of macroeconomics (economic theory); microeconomics (firm theory); and the new microeconomics, which, unlike traditional, studies economic processes at the molecular level of households, enterprises, and firms, is based on the atomic structure of society, and considers the only subject of decision-making individuals. Therefore, an important task of the researcher is to choose a specific quantitative apparatus, mathematical and statistical methods, and models for each level separately.

In the history of economics, there has been a constant debate about the relationship between qualitative and quantitative analyses. The proportion between them is slowly but steadily changing in favor of quantitative methods. Thus, the ideas of empirical science have significantly influenced the development of economics, contributing to the penetration of quantitative analysis, in particular, measurement, mathematical and statistical modeling, and forecasting.

It is the measurement or, rather, the impossibility of direct measurement on the metric scale of some economic features that prompted the scientific community to use in their analysis of multidimensional statistical methods, including taxonomic, discriminant, factor, fuzzy set theory, and fuzzy logic. Here, by such unmeasurable economic indicators, we do not mean purely qualitative features such as company name, gender, and profession, but complex hierarchical concepts, the level of which is judged only by experts using ordinal scale categories and which are manifested on the surface of economic reality.

Such hidden (latent) features at the company level include financial condition, competitiveness, investment attractiveness, sustainable development, labor efficiency of certain categories of workers (e.g., employees), as well as product competitiveness, the global economic criterion of the optimal production program, priority of domestic investment projects, etc.

A significant contribution to the development of economics was made by the outstanding American mathematician of Azerbaijani origin, L. Zadeh (1921–2017), who developed methods for fuzzy sets and fuzzy logic. The fact is that in recent decades, the assessment of latent traits in the economy has become a widely used method that uses in the processing, expert assessments of the concepts and terms of fuzzy logic, including fuzzy numbers, linguistic variables, membership functions, and more.

In his works [1–3], Zadeh laid the foundations of a new formal and mathematical apparatus, a new logic, which is as close as possible to the thinking and logic of man himself. The first stage in the formation of a new direction in mathematical theory began with the study of fuzzy set theory.

For example, at the second stage of development of this scientific direction, fuzzy logic and the possibility of its stagnation in practice, E. Mamdani published a study on the problem of regulating a power plant with a fuzzy steam generator [4] and proposed and implemented fuzzy controllers [5]. During this period, the development of fuzzy theory continues with sets and fuzzy logic [6], including the definition of operations on fuzzy numbers [7, 8]. At the same time, expert systems based on fuzzy logic began to develop rapidly, which were widely used in medicine and economics, particularly in finance [9].

Finally, in the third stage, the fuzzy approximation theorem (FAT) confirmed the completeness of fuzzy logic [10].

During this period, 48 Japanese companies formed a joint laboratory called LIFE (Laboratory for International Fuzzy Engineering). It facilitated the direct use of fuzzy logic algorithms in nonlinear control applications; in systems that are self-learning and change their structure; in studies of risk and critical situations; when recognizing images; in financial analysis (in the securities markets); in the study of panel data repositories; in the process of improving management strategies and coordination, for example in the management of complex large-scale production [11, 12].

It should be noted that the basic concepts and mathematical foundations of fuzzy set theory and fuzzy logic are considered in great detail in modern scientific literature [13–17]; so we will not dwell on them. The main attention will be paid in the future to the less studied section of economics — the theory of latent features, methods of their evaluation, and the role and place among them of fuzzy-plural apparatus, according to L. Zadeh.

1.2 The Concept and Essence of Latent Economic Characteristics

Problems of quantifying the level of the desired hidden properties of the studied objects, such as companies, have always been of interest to scientists and practitioners. The fact is that timely signals of negative trends in the development of such features as competitiveness, investment attractiveness, financial condition, and sustainable development allow taking appropriate measures in tactical crisis management and their constant monitoring is the basis for developing long-term sustainable development. The main purpose of assessing the latent features of economic objects is to obtain reliable information for management decisions. On the one hand, this enables the company's management to adjust the concept of development of this hidden property and change the strategy in this area. And, on the other hand, external users (investors, creditors, and public authorities) in the implementation of specific plans in relation to the company: mergers, acquisitions, investments, loans, etc. [18–22].

The concept of latent feature is closely related to the possibility of using certain scales in its measurement, which means a special procedure, that results in a numerical model of the studied features of the object.

Under the scale is understood as a sign system, for which a certain unambiguous mapping is given, which corresponds to the real objects of one or another element of the scale. The theory of measurement based on scaling was introduced by an American psychologist, S. Stevens [23]. Stevens put forward four types of such numerical systems, which led to the corresponding scales of measurement: nominal scale; ordinal scale; absolute scale; proportional scale.

The classification of economic characteristics, which is based on various scales of measurement, is presented in Figure 1.1.

Latent features in the literal sense of the word surround the researcher, especially at the micro-level. Examples of ordinal scales in the economy that reflect the relationship between the hidden properties of individual companies are ratings of the level of quality of products by different manufacturers, ranks of their financial condition, investment attractiveness, competitiveness, the priority of investment projects, and more.

In the process of developing the relevant field of knowledge, the type of scale for the same feature may change. With the development of modern economics and the use of mathematical and statistical models, including

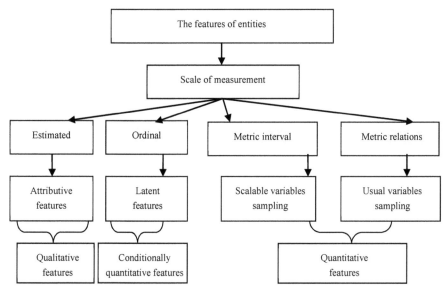

Figure 1.1 Classification of features of economic objects [18, 19, 23].

fuzzy set methods, some features may move from one scale to another (usually from left to right in Figure 1.1) in accordance with the process of approaching this field of exact science. The fact is that some latent features of economic objects can be estimated and presented in the form of scalable metrics on the interval scale. Due to certain circumstances, it is not possible to give an unambiguous quantitative description of these properties of economic objects, and their levels are usually judged by experts using gradations of the ordinal scale. The basis of such an assessment is the external manifestation of the hidden feature in the form of a set of values of the symptom factors X_1, $X_2, \ldots$, which characterize its various aspects. Figure 1.2 shows the model of the latent sign "competitiveness" companies, the basis of which is the assumption that the observed values of common factors (metric indicators scales X_1, X_2, etc.) (third level of the hierarchy) are an external manifestation of some hidden characteristics (first and second levels of the hierarchy) that cannot be measured in the metric scale.

For example, the latent production component at the company level reflects the following technical and economic indicators of the third level of the hierarchy in Figure 1.2, the capital of workers, the share of modern technologies in the overall technological cycle, the degree of computerization

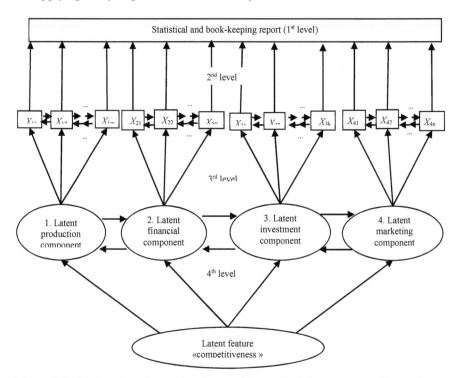

Figure 1.2   Model of the latent sign "competitiveness" of the company (ellipses denote hidden economic signs, and squares and rectangles denote metric scale indicators) [13, 14, 16].

of production and management, and material and energy consumption of products [24].

1.3 Review of Existing Methods for Assessing Latent Economic Characteristics

In recent years, the rapid development of research on latent features of economic objects [25] has led to the use of many techniques, algorithms, and procedures that are quite unequal in their purpose. The lack of clear systematization causes a further expansion of the tools for studying hidden properties, which are often inefficient and subjective [26, 27]. Therefore, it is important that at this stage of the development of science, basic classification of methods for assessing latent economic characteristics be created [28].

Figure 1.3 shows the classification of methods for assessing latent traits in economic facilities. Each level of detail is determined by its classification

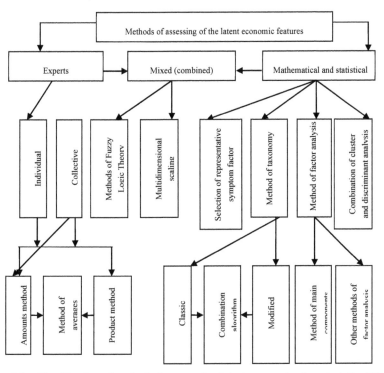

Figure 1.3 Classification of methods for assessing latent economic characteristics [12, 14].

feature: the degree of formalization of the source information, the general principle of application, and a variety of partial procedures and algorithms. According to the degree of formalization of the source information, all methods are divided into expert, mixed, mathematical, and statistical.

Expert methods of assessing latent traits are historically the first, traditional, and most popular. Their application began with researchers' scientific interest in economic features such as financial condition, competitiveness, investment attractiveness, efficiency, productivity, labor intensity, product competitiveness, and so on. At the same time, they usually turn to a survey of specialists in the relevant field of economics, who, based on their own experience, knowledge, and information about the magnitude of symptomatic factors, determine the priorities of the studied objects for certain hidden properties. The obtained individual and collective expert evaluations are used as final results or as initial data in combined methods of evaluation of latent features, for example, when using fuzzy set and fuzzy logic methods.

At the same time, it is hypothesized that experts can understand the importance of the task with a high degree of reliability and the feasibility of choosing from a set of possible estimates of a certain compromise option. Today, methods of collective expert evaluation of hidden properties of companies based on the work of special commissions are widespread, when groups of experts at a round table discuss the problem in order to agree and develop a single required assessment (or an acceptable range of assessments). This method, along with the subjectivism inherent in any expert evaluation, also has a disadvantage as a significant psychological pressure of collective opinion on individual professionals.

As a rule, the most important task in assessing latent traits is the processing of the results. When forming a group of experts, the main issues are its qualitative and quantitative compositions. In order to obtain a high-quality team of experts, a number of requirements are set for the group members, among which the following are the most important:

- high level of general economic erudition;
- deep special knowledge in the field of this hidden feature;
- availability of scientific interest or practical skills in relation to the assessed property of companies.

It is difficult to determine the optimal quantitative composition of the group of experts. At present, formalized approaches to the calculation of this value have been developed.

Based on mathematical formulas that give the possibility to assess the competence of a potential expert, the maximum and minimum limits of the group size, find the limits of the required quantitative composition of the group.

Expert estimates are obtained and are usually presented in the form of levels of digitized ordinal scale, for example, in points. Because the expert has less difficulty in answering the question as to which of the two companies surveyed is more attractive for investment than indicating their approximate level in any unit of measurement.

Following are the types of information used in these methods:

- opinion in the form of an appropriate number within the proposed limits (assessment of the latent feature of each company);
- ranking of companies (their placement);
- clustering of the whole set of companies into separate classes and groups;

- pairwise comparison of companies and reports on how they are "better" on the basis of the studied hidden feature.

Carrying out any examination includes the following main stages:

1. formation of the purpose of examination;
2. selection of experts;
3. choice of survey methodology;
4. processing of the received information, including a check of consistency and reliability of expert estimations.

However, it should be noted that the procedure of assessing latent features in the economy using expert methods has its specific features related to the complexity and versatility of such hidden properties as competitiveness, financial condition, investment attractiveness, sustainability, the priority of investment projects, and more. This leads to the implementation of, in addition to the four stages of examination, a number of additional stages, among which the most important are:

1. creation of the initial assessment environment, i.e., determination of a certain range of factors-symptoms (including hidden) of the studied latent feature;
2. establishing and taking into account the system of statistical weights that reflect the degree of importance of individual factors-symptoms;
3. scaling (normalization) of symptomatic factors, the main purpose of which is to avoid the erroneous influence of their units of measurement on the final results of the evaluation of the latent property under study. In the case of the use of latent symptom factors by rationing, the quantitative indicator is reduced to a fuzzy number in the range from 0 to 1;
4. direct assessment of the hidden property of economic objects – the calculation of the required, usually in points or in other conventional units, integrated (synthetic) indicator;
5. use of the obtained results – ranking of companies according to the level of the estimated latent feature, development of appropriate recommendations, development strategies, and so on.

At the first stage, the substantiation of a set of factors-symptoms of the latent property of the companies that are included in the further research is carried out. The future environment of indicators must meet the following conditions:

- Selected factors-symptoms reflect the most important from the point of view of the evaluated latent feature of the company, and their number is limited.
- Determining the factors-symptoms should focus on the use of public financial and statistical reporting data while minimizing the use of inside information.

It should be borne in mind that the definition of a range of factors-symptoms of latent traits of companies often turns into an independent scientific problem, the solution of which depends on numerous specific conditions and circumstances of the study – the field in which features of production and sales, etc.

Mathematical and statistical procedures and algorithms are relatively "young" methods for estimating latent features of economic objects, the emergence of which is associated with the development and implementation of multidimensional statistical and mathematical models, including correlation–regression, factor, cluster, discriminant, and taxonomic analysis, as well as multidimensional scaling. Despite the fact that the mathematical and statistical apparatus of these methods was developed long ago (for example, the foundations of correlation theory were laid by K. Pearson in the late nineteenth century) [29], their application in various fields of science and practice was hampered by a lack of technical means to solve complex high-dimensional problems. And only in the second half of the last century, with the development of computer technology, have these methods become widespread in the study of the hidden properties of any object, including economics.

One of the simplest methods of estimating the latent economic feature is to select the most representative vector of the matrix of initial data X. His idea is as follows: all the factors-symptoms (vectors-columns of the matrix X) are interconnected by a linear relationship of different degrees of density, i.e., multicollinearity. The more closely connected this column vector X_k is with all others, the more the reason to consider it as a linear combination of vectors $X_1, X_2, \ldots, X_m$ and, accordingly, as a carrier of information about the whole set of factors-symptoms.

The taxonomic analysis is one of the multidimensional procedures that allow you to get an idea of the magnitude of the studied latent trait. It is based on the calculation of distances (similarities) of all points (companies) to the standard in the space of the observed symptom factors, which are the external manifestation of the hidden properties of objects. This method is embodied in two main algorithms of taxonomic analysis – classical and modified. If the classical procedure involves determining the similarity with the standard, the modified procedure is based on the calculation of distances to the opposite of the standard.

It should be noted that the method of taxonomy is not completely free of possible errors: the classical algorithm sometimes leads to errors in determining the similarity with outsider objects (for the studied hidden property of objects), and the modified algorithm, on the contrary, suffers from inaccuracy in identifying leaders. At present, to eliminate or reduce the probability of these errors, a mixed algorithm of taxonomic analysis has been developed, which combines the positive aspects of the two main procedures.

The basis of factor analysis in the mathematical sense of the term as a method of assessing latent traits is the assumption that the observed values of factors-symptoms – metric scale X_1, X_2, ..., X_m – are an external manifestation of some latent property that characterizes economic objects. The task of factor analysis is to simulate the values of X_1, X_2, ..., X_m to quantify the latent characteristic using artificial variables F_1, F_2, ..., F_m, which are linear combinations of symptom factors.

The discussed situation can be commented on as follows: indicators of the metric scale X_1, X_2, ..., X_m of economic objects correlate with each other because they are under the influence of a hidden property. But if the factors-symptoms of objects are directly observed and measured, the magnitude of the latent feature can be judged only indirectly – by the size and closeness of the relationship of variables X_1, X_2, ..., X_m.

Among the algorithms of factor analysis, which are described in detail and comprehensively in the mathematical and statistical literature, we can indicate the following best-known methods that differ in the method of determining the factor decision:

- main components;
- maximum plausibility;
- centroids;
- main axes;
- approximate generalities.

It is proved and confirmed by practice that they all provide well-consistent results with a fairly high pairwise correlation coefficient. Therefore, usually use the most popular type of factor analysis – the method of principal components.

The combination of cluster and discriminant analysis in assessing the hidden properties of economic objects was first used in the 1960s of the twentieth century. In determining the probability of bankruptcy of Western companies using multifactor models. The task of cluster analysis is to divide the set of studied objects into homogeneous groups according to the level of the latent feature under study. This grouping is based on the values of its factors-symptoms $X_1, X_2, \ldots, X_m$ using known algorithms of cluster analysis — agglomerating hierarchical procedure, methods of k-averages, "trout", "optimization algorithms," and others. Further, on the basis of the method of double unification (both by objects and factors-symptoms), the identification of selected clusters is carried out, i.e., the establishment of groups of objects-leaders, objects-middlemen, objects-outsiders, etc.

In the process of discriminant analysis, the following main tasks are solved:

- construction of an adequate discriminant function using correlation and regression methods;
- determination of estimates of the latent feature of the studied economic objects on the basis of the constructed discriminant function;
- evaluation of the hidden property under study for new objects, i.e., those that are not included in the studied population.

The first problem is solved by calculation and statistical analysis of the usual regression equation, in which the role of the performance indicator is performed by an artificial variable that determines the affiliation of each economic object to a particular cluster. And as factor indicators are factors-symptoms $X_1, X_2, \ldots, X_m$. In this case, the accuracy of future estimates of the latent property of objects can be determined on the basis of the value of the coefficient of determination of the constructed adequate discriminant function in the form of the found regression equation.

The second task is to calculate the required estimate of the latent feature under study for each object of the studied population according to the known values of factors-symptoms $X_1, X_2, \ldots, X_m$ on the basis of the discriminant function.

The third task is the direct task of discriminating against new economic objects, i.e., predicting the level of hidden properties of objects that are not part of the studied population, and assigning them to a particular cluster.

Mixed (combined) methods for estimating the latent properties of objects are based on a combination of both expert and mathematical−statistical procedures and algorithms. These include methods of fuzzy logic theory, and metric and nonmetric scaling. Combined methods usually have the same advantages and disadvantages as expert and mathematical–statistical approaches to assessing the latent characteristics of economic objects.

Methods of multidimensional scaling, in particular, one of their most popular areas (nonmetric scaling), consists of two consecutive mathematical and statistical procedures: cluster and factor analysis. In this case, the input obtained is the information obtained from the expert survey, which is measured on the ordinal (rank) scale.

1.4 Estimation of Latent Economic Features based on a Combination of Expert Methods and Methods of Fuzzy Sets and Fuzzy Logic

Usually, the process of expert evaluation of latent features at the level of the manufacturing company is two-stage and goes in the opposite direction compared to Figure 1.2. First, the levels of factors-symptoms $X_1, X_2, \ldots X_2$ are analyzed, which are measured in the metric scale and directly observed, or calculated based on statistical reporting (third and fourth levels of the hierarchy in Figure 1.2). They are determined by specialists in the field of this hidden property (production, financial, investment, marketing, and other components), studied, and modeled using fuzzy set theory and fuzzy logic up to group latent features (first stage) based on the estimated components (second level of the hierarchy) to the most desired latent feature (first level of the hierarchy in Figure 1.2).

However, in certain circumstances, such as the need for rapid assessment of latent symptoms and rapid intervention in the development of the production system, this process may be one-step. In this case, only the first and second levels of the hierarchy in Figure 1.2, show that experts, based on their knowledge and experience, use the methods of fuzzy sets and fuzzy logic to evaluate both the hidden group and final characteristics of companies.

One of the most important concepts in the expert evaluation of latent economic features is the concept of linguistic variables. By definition, these are variables that cannot be described by mathematics, i.e., it is difficult to provide an accurate (objective) quantitative estimate. For example, the concepts of "small" or "medium" (business) and "high" or "low" (interest rate) do not have a clear limit and cannot be represented by an accurate mathematical description [2, 6].

The classical linguistic variable includes five main components: (a, T, X, G, M), where a is the name of the linguistic variable and T is the basic term set of a linguistic variable, or its many meanings — terms (from the English term — to call). Each term reflects a set of subjective characteristics of the expert and is usually not numerical; X is the domain (universal set) of fuzzy variables included in the term set; M is the syntactic rule, which generates new values of the linguistic variable from the basic term set T; M is a semantic rule obtained by the rule G, some fuzzy sets.

For example, when estimating the latent variable "competitiveness" of the company (Figure 1.2) for a linguistic variable called "Level of the financial component of the company" (component a), the term set has the following form: $T = \{$"very low," "low," "medium," "high," "very high"$\}$. The area of the definition of X is the range of changes in the financial component of the company in points (0, 5). These rules are reduced to the formation of membership functions $\mu_T(X)$, taking into account logical connections and modifiers. It should be borne in mind that linguistic variables in addition to verbal meanings can sometimes be numerical.

Consider in more detail the essence of the one-stage procedure of expert evaluation of the latent feature "Level of competitiveness" and group hidden features (first and second levels of the hierarchy in Figure 1.2) for a set of n companies (i — company number, $i = 1, 2, \ldots, N$) based on their combination with the methods of fuzzy set theory and fuzzy logic.

We introduce additional notation: h is the number of the group latent feature ($h = 1, 2, 3, 4$, etc.). For each of these latent symptom factors, the linguistic variable "Level of symptom factor X_h" is introduced with the term set of values of T_{sh} ($s = 1, 2, \ldots, 5$), each of which is a fuzzy set described in natural language: T_{1h} is "Very low level of symptom factor X_h," T_{2h} is "Low level of symptom factor X_h", T_{3h} is "Average level of symptom factor X_h," T_{4h} is "High level of symptom factor X_h," and T_{5h} is "Very high level of symptom factor X_h." This approach to building an economic–mathematical model corresponds to a five-point system for assessing group latent factors-symptoms: production, financial, investment, and marketing components of the company's competitiveness [30, 31].

Suppose that for one of them, for example, the production one ($h = 1$), the correspondence of the term sets T_{sh} to the trapezoidal fuzzy numbers is established (Table 1.1).

Correspondence tables are similarly constructed for the other three hidden components of the company's competitiveness ($h = 2, 3, 4$, etc.). Next, the fuzzy correspondence of the current value of the indicator of the production component *and* the research company to each linguistic value of the corresponding term set is determined. The functions of belonging of the values of

Table 1.1 Correspondence of term sets to trapezoidal fuzzy numbers.

Fuzzy set	Fuzzy estimate in the form of a trapezoidal number
T_{1h}	(0.0, 0.0, 0.15, 0.25)
T_{2h}	(0.15, 0.25, 0.35, 0.45)
T_{3h}	(0.35, 0.45, 0.55, 0.65)
T_{4h}	(0.55, 0.65, 0.75, 0.85)
T_{5h}	(0.75, 0.85, 0.95, 1.0)

the indicator of the production component to fuzzy sets are set. In this case, the membership functions must satisfy the following conditions:

- Carriers of any three linguistic values of the financial component do not intersect, and carriers of two consecutive linguistic values of the financial component intersect.
- If the current value of the financial component belongs to the carrier of one linguistic value, the membership function is equal to 1; if the current value of the indicator of the financial component belongs to the intersection of the carriers of two consecutive linguistic values, then the sum of the membership functions of the corresponding fuzzy sets is also equal to 1.

Such properties are satisfied, for example, by the system of trapezoidal membership functions $\mu(X)$, which are often used in practical research using elements of fuzzy set theory and fuzzy logic.

Figure 1.4 shows a graph of the membership function of the scale of the production component in the form of trapezoidal functions, which is based on the data in Table 1.1. Similarly, the functions belonging to the score scale are built according to expert evaluation and their graphs of other latent factors-symptoms to obtain a fuzzy classification of all the hidden components of the company's competitiveness.

As can be seen from Figure 1.4, fuzzy numbers are engaged one after another. This reflects the fact that there is no sharp division between adjacent estimates, and the transition from one term set to another is gradual. The result of the evaluation of group latent features X_h, however, as well as the competitiveness of the company, is a fuzzy number lying on the interval from 0 to 1.

Then move on to ranking the hidden components by their levels based on expert assessments. For this purpose, the weights of the symptom factors X_h are determined, i.e., the weights w_h are calculated. Finding the weight for indicators is the most important and meaningful stage of the procedure. At

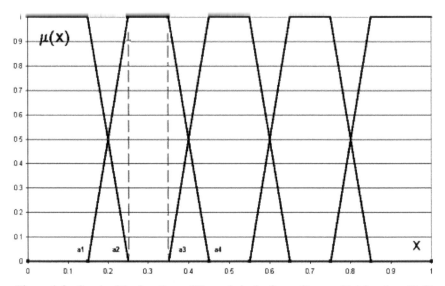

Figure 1.4   Graph of the functions of the scale in the form of trapezoidal functions [7, 8].

this stage, the researcher decides which factors-symptoms (including latent) are of the highest priority over others, which ultimately determines the type of the solution sought.

The generally accepted method of constructing the weights of symptomatic factors is the analytical hierarchical method proposed by T. Saaty [30], which consists in the following for each pair of indicators with numbers f and g; experts estimate the number w_{fg}, which shows how many times the first factor exceeds the second. It is considered that in an ideal situation, equality holds: $W_{fg} = \frac{w_f}{w_g}$, where w_f and w_g are the weights of the symptom factors f and g, respectively. T. Saaty [32] developed a comparison algorithm that allows us to find a set of weights by the coefficients w_g, w_f, and w_g.

The following pairwise comparisons are represented by the matrix in eqn (1.1):

$$
W = \begin{array}{c} \\ X_1 \\ X_2 \\ X_1 \end{array}
\begin{array}{c} X_1 \quad X_2 \quad X_h \\ \left(\begin{array}{ccc} w_{11} & w_{12} & w_{1h} \\ w_{21} & w_{21} & w_{2h} \\ \dots & \dots & \dots \\ w_{h1} & w_{h2} & w_{hh} \end{array} \right) \end{array}.
\tag{1.1}
$$

The matrix of pairwise comparisons (1.1) is diagonal and inversely symmetric. In the research works [32] and [33], Saaty's method was extended to the case of fuzzy sets. The membership functions are assumed to be equal to the corresponding coordinates of the eigenvector $V = (v_1, v_2, \ldots, v_h)^T$ of the matrix of paired comparisons W:

$$X h \ v h \, , h \ 1, \ 2, \ 3, \ 4. \tag{1.2}$$

The eigenvector V is from the following system of requirements:

$$WV = \max \begin{cases} \max V \\ v1 + v2 + \cdots + vh = 1 \end{cases}, \tag{1.3}$$

where λ_{max} is the maximum eigenvalue of the matrix W.

The ranking results are presented in the table 1.1, the subject of which is n of the studied companies, and in the predicate — the levels of the term set T_{sh}. The values of the ranks λ_{ish} are calculated in this way to meet the following conditions:

$$0 \leq_{ish} \leq 1; \ \sum_{s=1}^{5} ish = 1 \tag{1.4}$$

for any values of i and s.

Using the obtained coefficients w_h, find for each company the average weighted latent factors-symptoms X_h ranks:

$$ish = \sum_{h=1}^{4} w_{hish}, \tag{1.5}$$

where λ_{is} the weighted average rank of the sth level of the term set for the ith company. λ_{is} is calculated in this way to satisfy condition (1.4) for any values of i.

Direct construction of the integrated assessment of the studied hidden property of the first level of the hierarchy in Figure 1.2 begins with the introduction of the linguistic variable "Company competitiveness level" with a term set of values "Very low level of company competitiveness," "Low level of company competitiveness," "Medium level of company competitiveness," "High level of company competitiveness," "Very high level of competitiveness," and "Companies." The carrier of its term set is determined – the real variable U in the interval from zero to one. A system of trapezoidal

Table 1.2   Values membership functions for a linguistic variable "The level of competitive ness of the company."

The value of U	Very low	Low	Average	High	Very high
0−0.15	1	0	0	0	0
0.15−0.25	$(0.25-U)/0.1$	$(U-0.15)/0.1$	0	0	0
0.25−0.35	0	1	0	0	0
0.35−0.45	0	$(0.45-U)/0.1$	$(U-0.35)/0.1$	0	0
0.45−0.55	0	0	1	0	0
0.55−0.65	0	0	$(0.65-U)/0.1$	$(U-0.55)/0.1$	0
0.65−0.75	0	0	0	1	0
0.75−0.85	0	0	0	$(0.85-U)/0.1$	$(U-0.75)/0.1$
0.85−1	0	0	0	0	1

membership functions of the corresponding fuzzy sets is given, the domain of which will be the interval from zero to one (Table 1.2).

Integral assessment of the latent feature "competitiveness" U for each ith company is calculated by the following formula:

$$U_i = \sum_{s=1}^{5} \alpha_{sis} , \qquad (1.6)$$

where α_s are the abscissas of the maxima of the membership functions of the term sets of the linguistic variable.

For trapezoidal membership functions, the value of α_s is according to the following formula, provided that the corresponding trapezoids are equilateral:

$$as = 0.2 \times s - 0.1. \qquad (1.7)$$

Further, according to Table 1.1, the level of fuzzy required latent features in terms of linguistic variables of each company is calculated from the totality of all objects.

In solving specific practical problems of estimating the latent economic feature using the basic elements of fuzzy set theory and fuzzy logic, it is recommended to use mathematical software [34, 35].

1.5 Conclusion

In our opinion, the discussed approach essentially represents further development and formalization of ideas of a method of matrix models as one of several kinds of expert rating estimation. Its improvement is that the estimates provided by experts for all latent traits (both factorial and productive) are strictly scalable and belong to the interval [0, 1]. And some of them may intersect, forming a fuzzy set. Thus, the main advantage of integrated indicators constructed with the help of elements of the fuzzy logic theory, which are considered as estimates of the studied latent features, is the formalization of opinions of experts in the process of rating assessment based on matrix models. However, we should not forget about all the negative features of expert evaluation (a certain subjectivity, the problem of creating competent groups of experts, etc.).

A significant shortcoming, which is manifested each time in the application of elements of fuzzy set theory and fuzzy logic, is that when performing calculations using a fuzzy set approach, the results of calculations largely depend on the correct division of a fuzzy set at a level that is sometimes quite difficult to do with limited and rather heterogeneous statistical information.

Nevertheless, we hope that this area of science, which is at the intersection of economics and mathematics, will continue to develop successfully, attracting new researchers to the ranks of its followers and revealing the hidden properties of economic objects. And the list of the latter will be further expanded due to the penetration of ideas and methods of fuzzy evaluation of latent traits in all areas of the economy.

References

[1] L. A. Zadeh, 'Fuzzy sets'. Inf. Control 1965, 8, 338–353.

[2] L. A. Zadeh, 'Is there a need for fuzzy logic?' Inf. Sci. 2008, 178, 2751–2779.

[3] L. A. Zadeh, 'Fuzzy Sets as a Basis for a Theory of Possibility. Fuzzy Sets and Systems, 1978, 1, 3–28.

[4] E. H. Mamdani, 'Application of fuzzy algorithms for control of simple dynamic plant', Proceedings of the Institution of Electrical Engineers, vol. 121, no. 12. Institution of Engineering and Technology (IET), p. 1585, 1974. doi: 10.1049/piee.1974.0328.

[5] E. H. Mamdani and S. Assilian, 'An experiment in linguistic synthesis with a fuzzy logic controller,' International Journal of Man-Machine Studies, vol. 7, no. 1. Elsevier BV, pp. 1–13, Jan-1975.

[6] R. E. Bellman and L. A. Zadeh, 'Local and Fuzzy Logics,' Modern Uses of Multiple-Valued Logic. Springer Netherlands, pp. 103–165, 1977.

[7] D. Dubois and H. Prade, 'Operations on fuzzy numbers,' International Journal of Systems Science, 9(6), pp. 613–626, 1978.

[8] D. Dubois and H. Prade, 'Fuzzy sets and systems: theory and applications.' Academic Press, Boston, 1980.

[9] J. J. Buckley, 'The fuzzy mathematics of finance', Fuzzy Sets and Systems, 21, pp. 257–273, 1987.

[10] B. Kosko, 'Fuzzy Thinking: The New Science of Fuzzy Logic'. Hyperion, Disney Books, 1993.

[11] G. J. Klir, 'Foundations of Fuzzy Set Theory and Fuzzy Logic: A Historical Overview', International Journal of General Systems, 30, pp. 91–132, Jan-2001.

[12] L. A. Zadeh, 'Fuzzy logic—a personal perspective', Fuzzy Sets and Systems, 281, pp. 4–20, Dec-2015.

[13] F. Dernoncourt, 'Introduction to fuzzy logic'. Massachusetts Institute of Technology, 21, 2013

[14] H. J. Zimmermann, 'Fuzzy set theory—and its applications'. Springer Science & Business Media, 2011.

[15] J. J. Buckley and E. Eslami, 'An Introduction to Fuzzy Logic and Fuzzy Sets'. Physica-Verlag HD, 2002.

[16] S. Gottwald, 'Fuzzy sets and fuzzy logic: The foundations of application—from a mathematical point of view'. Springer-Verlag, 2013.

[17] Y. Bai, D. Wang, 'Fundamentals of fuzzy logic control—fuzzy sets, fuzzy rules, and defuzzifications'. In Advanced fuzzy logic technologies in industrial applications (pp. 17-36). Springer, London, 2006.

[18] A. G. Yankovyi, 'Factor analysis and the method of principal components in the study of latent indicators'. Market economy: Modern theory and practice of management, vol. 6, no. 7. 2003, pp. 81–104.

[19] O. G. Yankovyi, O. L. Gura, 'Methods and models for assessing latent indicators in business planning'. Bulletin of social and economic achievements of ODEU, 2006, 24, pp. 399–403.

[20] O. Yankovyi, Yu. Goncharov, V. Koval, T. Lositska, 'Optimization of the capital-labor ratio on the basis of production functions in the economic model of production'. Naukovyi Visnyk Natsionalnoho Hirnychoho Universytetu, 4., Aug-2019.

[21] O. Yankovyi, 'Latent features in economics'. Odesa, Atlant, 2015.

[22] O. Yankovyi, V. Koval, L. Lazorenko, O. Poberezhets, M. Novikova, V. Gonchar, 'Modeling Sustainable Economic Development Using Production Functions'. Studies of Applied Economics, 39(5), 2021.

[23] S. S. Stevens, 'Mathematics, measurement, and psychophysics'. Handbook of experimental psychology, 1951.

[24] O. Borodina, H. Kryshtal, M. Hakova, T. Neboha, P. Olczak, V. Koval, 'A conceptual analytical model for the decentralized energy-efficiency management of the national economy'. Polityka Energetyczna – Energy Policy Journal, 25(1), 5-22, 2022.

[25] G. Chen, T. T. Pham, Introduction to fuzzy sets, fuzzy logic, and fuzzy control systems. CRC Press, 2000.

[26] L. Běhounek, P. Cintula, 'Fuzzy class theory'. Fuzzy Sets and Systems, 154(1), pp. 34–55, 2005.

[27] H. T. Nguyen, B. Wu, 'Fundamentals of statistics with fuzzy data'. New York: Springer, 2006.

[28] E. D. Cox, Fuzzy logic for business and industry. Charles River Media, Inc., 1995.

[29] H. E. Soper, a. W. Young, B. M. Cave, A. Lee, K. Pearson, "On the distribution of the correlation coefficient in small samples. Appendix ii to the papers of 'student' and R. A. Fisher. A cooperative study'. Biometrika, 11(4), pp. 328–413, 1917.

[30] S. E. Pogodayev, 'Marketing of works as a source of the new hybrid offerings in widened marketing of goods, works and services'. Journal of Business & Industrial Marketing, 28, pp. 638–648, 2013.

[31] O. Hutsaliuk, V. Koval, O. Tsimoshynska, M. Koval, H. Skyba, 'Risk Management of Forming Enterprises Integration Corporate Strategy'. TEM Journal, 9(4), pp. 1514–1523, 2020.

[32] T. Saaty, 'How to make a decision: The Analytic Hierarchy Process'. European Journal of Operational Research, 48, pp. 9–26, 1990.

[33] D. Y Chan, 'Application of extent analysis method in fuzzy AHP'. European Journal of Operation Research, 95, pp. 649–655, 1996.

[34] J. W. Lee, S. H. Kim, 'Using analytic network process and goal programming for interdependent information system project selection'. Computers & Operations Research, 27, pp. 367–382, 2000.

[35] F. Lefley, J. Sarkis, 'Applying the FAP model to the evaluation of strategic information technology projects'. International Journal of Enterprise Information Systems, 1, pp. 69–90, 2005.

2

A Fuzzy-based Group Decision-making Approach for Supplier Selection

Balaram Dey, Bipradas Bairagi, and Goutam Bose

Department of Mechanical Engineering, Haldia Institute of Technology, India

Abstract

Nowadays, proper supplier selection is a very important strategic concern in any supply chain management. Due to vagueness, ambiguity, and imprecise information regarding alternative suppliers with a constantly varying nature of selection criteria, proper decision-making becomes challenging. This chapter applies a fuzzy multiple-criteria homogeneous group decision-making process for performance evaluation and selection of suppliers in supply chain management (SCM). In this method, a homogeneous decision-making committee is formed to assess the alternatives and the criteria weight in terms of raw linguistic expression. These linguistic expressions are converted into equivalent triangular fuzzy numbers. Normalization of performance ratings is carried out, and a weighted normalized performance rating is determined by combining the rating and their respective weights. Fuzzy multi-objective optimization by ratio analysis (FMOORA) and fuzzy gray theory (FGT) methods are independently employed to calculate the net score and gray relation grade for each alternative to provide appropriate decision-making. A suitable numerical example is chosen for illustration of the approach. A comparative study on the results is conducted, and the best supplier is selected. The study indicates the applicability and effectiveness of the approach enunciated in the chapter.

Keywords: FMOORA, FGT, group decision-making, Supplier selection.

2.1 Introduction

Business supply chain faces tough, complex, and cut-throat competition in sustaining their existence in today's dynamic industrial scenario [1]. The overall performance of an industrial supply chain fundamentally depends upon its various parameters, namely available facilities, type of transportation, inventories, and quality of information [2]. Supplier selection certainly plays a pivotal role in enhancing the performance of that chain [3]. The process of purchasing raw materials, parts, components, semi-finished parts, etc., gets a boosting impact through proper supplier selection. Thus, supply chain management (SCM) of any business entity draws large attention to that supplier selection process and consequentially gets major research interest from industry as well as academia [4-6]. Persistent efforts by researchers are always evolving suitable processes in finding better sourcing for raw materials and parts from outside markets [7]. Through effective supplier selection, management would be able to reduce the purchasing cost to a large extent and improve the profitability of the organization [8]. A proper decision-making framework in supplier selection upstream could certainly benefit the management of a supply chain toward that survival.

Supplier selection depends upon several criteria often conflicting with each other, which necessitates a well-structured multiple-criteria decision-making (MCDM) environment [9]. Given the utmost consequence of supplier/vendor selection, the decision-making process often comprises many experts of the organization and leads toward a balanced and efficient group decision-making (GDM) process [10, 11]. This chapter considers the decision-making process as homogeneous since experts in the diversified field have a common view in making a decision [12]. The complexity and paucity of supplier source information compel the researchers to apply fuzzy set theory in MCDM in dealing with decision-making in an uncertain environment [13]. This chapter tries to frame a fuzzy homogeneous group decision-making process to address the problems of supplier selection under uncertainty and severe complexity in a business supply chain.

Many renowned researchers paid their academic attention to the vendor/supplier selection process in a supply chain. Chen [14] and Kahraman *et al.* [15] utilized fuzzy group decision-making methods in making decisions in a complex industrial environment. Shyur and Shih [16] and De Boer *et al.* [17] utilized different MCDM methods for the evaluation and selection of supplier selection in diversified fields of a business environment. Talluri and Narsimhan [18] applied MCDM approaches in dealing with supplier selection in a supply chain. Huixia and Tao [19] used the gray system theory to solve vendor selection problems. Li *et al.* [20] utilized a gray-theory-based decision-making approach to the supplier selection problem. Brauers and Zavadskas [21] utilized the MOORA method in decision-making for a transition economy. Dey *et al.* [22] used the FMOORA methodology for making strategic decisions in a supply chain. Chakraborty [23] utilized the MOORA approach as a decision-making aid in manufacturing surroundings. Bairagi *et al.* [24] used a novel multi-criteria decision-making technique for the performance evaluation of material handling devices in the industry. Dey *et al.* [25] applied a new method MOPA for making an accurate decision in an industrial scenario. Dey *et al.* [26] applied a multi-member decision-making process in a supply chain based on group heterogeneity. Ozcan *et al.* [27] performed a comparative analysis among many MCDM methods and implemented them in management decision-making.

2.2 Fuzzy MCDM Techniques for Supplier Selection

Many researchers pioneered and utilized FMCDM approaches in the assessment and choice of suppliers in a highly complex and uncertain business environment. Few MCDM methods such as FMOORA and FGT are described in Sections 2.2.1 and 2.2.2 of this chapter. The stepwise algorithm of these two methods is written below.

2.2.1 Fuzzy MOORA method

Brauers [28] introduced the MOORA method. The idea of combining fuzzy set theory with the conventional MOORA method is termed FMOORA. The process comprises the following steps as described below.

- Step 1: Construction of decision matrix. The FMOORA method commences with the formation of a decision matrix consisting of a subjective measure of alternatives in linguistic terms as follows:

$$
V = \begin{array}{c} A_1 \\ \vdots \\ A_i \\ \vdots \\ A_m \end{array} \left[\begin{array}{ccccc} v_{11} & \cdots & v_{1j} & \cdots & v_{1n} \\ \vdots & \cdots & \vdots & \cdots & \vdots \\ v_{i1} & \cdots & v_{ij} & \cdots & v_{in} \\ \vdots & \cdots & \vdots & \cdots & \vdots \\ v_{m1} & \cdots & v_{mj} & \cdots & v_{mn} \end{array} \right], \qquad (2.1)
$$

where v_{ij} represents the linguistic term (variable) assessed for alternative i regarding criteria j. In the matrix, m and n indicate the alternative number and the number of criteria, respectively.

- Step 2: Transform linguistic terms into fuzzy numbers.

$$
X = \begin{array}{c} A_1 \\ \vdots \\ A_i \\ \vdots \\ A_m \end{array} \left[\begin{array}{ccccc} \tilde{x}_{11} & \cdots & \tilde{x}_{1j} & \cdots & \tilde{x}_{1n} \\ \vdots & \cdots & \vdots & \cdots & \vdots \\ \tilde{x}_{i1} & \cdots & \tilde{x}_{ij} & \cdots & \tilde{x}_{in} \\ \vdots & \cdots & \vdots & \cdots & \vdots \\ \tilde{x}_{m1} & \cdots & \tilde{x}_{mj} & \cdots & \tilde{x}_{mn} \end{array} \right], \tag{2.2}
$$

where $\tilde{x}_{ij} = (a_{ij}, b_{ij}, c_{ij})$ denotes the fuzzy number corresponding to v_{ij}.

- Step 3: Normalization of the ratings of alternatives. In FMOORA, a ratio system is employed. In this ratio system, a comparison is made between every estimated rating of alternatives with a certain denominator, which acts as a representative for every alternative. The following equation is recommended for carrying out the normalization process:

$$
\tilde{r}_{ij} = \left(\frac{a_{ij}}{\max(c_j)}, \frac{b_{ij}}{\max(c_j)}, \frac{c_{ij}}{\max(c_j)} \right), \quad j = 1...n, \tag{2.3}
$$

where $\tilde{r}_{ij}$ represents normalized performance rating (in TFN), and fuzzy of alternative i regarding criterion j.

- Step 4: Estimate the fuzzy weights of the criteria.

$$
\tilde{W} = \begin{bmatrix} \tilde{w}_1 & \cdots & \tilde{w}_j & \cdots & \tilde{w}_n \end{bmatrix}, \tilde{w}_j \text{ is a fuzzy number.} \tag{2.4}
$$

- Step 5: Computations of a fuzzy net score of the alternatives. The fuzzy net score for every alternative is calculated by the difference between the total weighted fuzzy score under benefit criteria and the total weighted fuzzy score under cost criteria. The following equation can be used to calculate the fuzzy net score for every alternative:

$$
\tilde{y}_i = \sum_{j \in B} \tilde{w}_j \tilde{r}_{ij} - \sum_{j \in C} \tilde{w}_j \tilde{r}_{ij}. \tag{2.5}
$$

In most of the real-life problems different weights are given to the attributes as per their relative importance. When the weights of attributes are taken into consideration, then eqn (2.3) can be expressed as follows.

$\tilde{y}_i = (\alpha_{ij}, \beta_{ij}, \gamma_{ij})$ is the fuzzy net score computed for ith alternative. $\tilde{w}_j$ denotes the weightage of jth criterion. B and C indicate benefit and cost criteria, respectively.

- Step 6: Best non-fuzzy performance (BNP) is the defuzzified value of the fuzzy net score of every alternative. The following equation is recommended for computing the BNP value:

$$\text{BNP}_i = \left[(\text{UP}_i - \text{LP}_i) + (\text{MP}_i - \text{LP}_i) \right] / 3 + \text{LP}_i. \qquad (2.6)$$

Here, LP, MP, and UP imply lower, middle, and upper points of the fuzzy net score, respectively.

- Step 7: The best alternative is one that is associated with the maximum BNP value and the worst alternative is one that possesses the minimum BNP value. BNP value may be positive, negative, or zero. Therefore, select the alternative with maximum BNP as the most suitable alternative.

2.2.2 Fuzzy gray theory (FGT)

The fuzzy gray theory is a well-organized method for the analysis of mathematical systems specified by vague information and inadequate facts under a fuzzy environment [29–31]. Grey theory has its advantages over fuzzy theory in judging the degree of fuzziness under an uncertain environment. In essence, the gray theory has the liberty in fixing the vague condition. The procedural steps with various normalization techniques used for a different sense of the criteria are furnished below.

- Step 1: Determine performance values $\tilde{x}_i = (\tilde{x}_{i1}, \tilde{x}_{i2}, ..., \tilde{x}_{in})$ for every alternative regarding every criterion. Here, $i = 1, 2, ..., m$ and $j = 1, 2, ..., n$. m denotes an alternative number, and n denotes criteria number.
- Step 2: Set the reference data $\tilde{x}_0 = (\tilde{x}_{01}, \tilde{x}_{02}, ..., \tilde{x}_{0n})$. Reference data consists of the most favorable value or target value for each criterion.
- Step 3: Accomplish normalization of performance values by using the following equations:

$$\tilde{r}_{ij} = \frac{\tilde{x}_{ij} - (\tilde{x}_j)_{\min}}{(\tilde{x}_j)_{\max} - (\tilde{x}_j)_{\min}}, j \in B \qquad (2.7)$$

$$\tilde{r}_{ij} = \frac{(\tilde{x}_j)_{\max} - \tilde{x}_{ij}}{(\tilde{x}_j)_{\max} - (\tilde{x}_j)_{\min}}, j \in C \qquad (2.8)$$

$\tilde{r}_{ij} = 1 - \dfrac{|\tilde{x}_{ij} - \tilde{x}_j^*|}{\max|\tilde{x}_{ij} - \tilde{x}_j^*|}$. optimum value x_j^*.

- Step 4: Compute the differences between data sets by the following equation:

$$\tilde{\Delta}_i = \{|\tilde{x}_{01} - \tilde{x}_{i1}|, |\tilde{x}_{02} - \tilde{x}_{i2}|, ..., |\tilde{x}_{0n} - \tilde{x}_{in}|\}. \qquad (2.9)$$

- Step 5: Defuzzify the fuzzy differences by $\Delta_i = \left\{ y_i : y_i = \dfrac{l_i + m_i + u_i}{3} \right\}$ TFN.

  Also, determine the global maximum $(\Delta_j)_{\max}$ and global minimum $(\Delta_j)_{\min}$ for each criterion.

- Step 6: Evaluate the gray relation coefficient by the following equation:

$$\lambda_i(j) = \dfrac{(\Delta_j)_{\min} + \mu(\Delta_i)_{\max}}{\Delta_i(j) + \mu(\Delta_j)_{\max}}. \qquad (2.10)$$

The value of the coefficient μ is in the interval $0 \le \mu \le 1$, but, usually, 0.5 is taken as the value μ. The coefficient μ is used to mitigate the effect of the global maximum value $(\Delta_j)_{\max}$.

- Step 7: Compute the gray relational grade (GRG_i) by means of the following formula:

$$GRG_i = \sum_{j=1}^{n} w_j \otimes \lambda_i(j), \qquad (2.11)$$

where w_j denotes the weightage of the jth criterion in fuzzy number. The alternative with the highest assessment of objective gray relational grade (GRG_i) is the best solution and the alternative with the least assessment of objective gray relational grade (GRG_i) is the worst solution.

Now, both these FMCDM methods are illustrated with a numerical example to show their applicability in Section 2.3.

2.3 Illustrative Example

For the purpose of demonstrating the employability of the FMOORA and FGT approaches, the authors cited a supplier selection problem from Wu and Liu [32]. Wu and Liu [32] simplified the said problem statement using information from Li *et al.* [20]. In this problem, four suppliers (S_1, S_2, S_3, and S_4) were initially considered as available alternatives. A set of five important evaluation criteria (C_1, C_2, C_3, C_4, and C_5) is also considered

Table 2.1   Linguistic terms, acronyms, and related triangular fuzzy numbers (TFN).

Linguistic terms	Acronyms	TFN
Absolute good	AG	(0.9, 0.95, 1)
Very good	VG	(0.8, 0.85, 0.9)
Good	G	(0.7, 0.75, 0.8)
Fairly good	FG	(0.6, 0.65, 0.7)
Medium good	MG	(0.5, 0.55, 0.6)
Medium	M	(0.4, 0.45, 0.5)
Fairly poor	FP	(0.3, 0.35, 0.4)
Poor	P	(0.2, 0.25, 0.3)
Very poor	VP	(0.1, 0.15, 0.2)
Absolutely poor	AP	(0, 0.05, 0.1)

for the decision-making process. The five criteria listed are reputation (C_1), service quality (C_2), product quality (C_3), delivery time (C_4), and price (C_5). Among all the criteria, C_1, C_2, and C_3 are the beneficial criteria; hence, the higher, the better. Contrarily, C_4 and C_5 are non-beneficial criteria; hence, the lower, the better. A decision-making committee comprising a few domain experts suggested the ratings of the suppliers in linguistic expressions and their equivalent triangular fuzzy numbers (TFNs) are shown in Table 2.1. Decision makers in that group having equal rights in deliberation and making decisions reflect the policy of the organization. The committee unanimously suggested the performance ratings of the alternative suppliers as shown in Table 2.2. The decision matrix of the alternatives is shown in Table 2.3. The fuzzy weights of the selection criteria are shown in Table 2.4. Now, FMOORA and FGT approaches are applied to the cited problem to evaluate and select the best supplier from among the available alternatives.

2.3.1 Illustration of the cited numerical with the FMOORA method

In Section 2.3.1, the FMOORA approach is utilized to solve the cited example. The fuzzy weighted decision matrix of the suppliers under FMOORA is shown in Table 2.5. The sum of the weighted fuzzy score of the suppliers under FMOORA, net score, and their ranking is shown in Tables 2.6 and 2.7. The downward order of net score values decides the ranking order of the suppliers. Here, supplier S_2 is adjudged as the best supplier with the highest net score (2.95). Similarly, supplier S_1 is considered the last choice with the least net score value (0.48). Accordingly, the ultimate ranking preference of the suppliers happened to be $S_2 > S_3 > S_4 > S_1$, as shown in Table 2.7.

Table 2.2   Decision matrix in linguistic terms [33].

Suppliers	C_1: Reputation (+)	C_2: Service quality (+)	C_3: Product quality (+)	C_4: Delivery time (−)	C_5: Price (−)
S_1	FG	M	FG	VG	VG
S_2	VG	FG	MG	M	MG
S_3	G	VG	MG	G	FG
S_4	VG	G	M	MG	VG

Table 2.3   Fuzzy decision matrix.

Suppliers	C_1 (+)	C_2 (+)	C_3 (+)	C_4 (−)	C_5 (−)
S_1	(0.6, 0.65, 0.7)	(0.4, 0.45, 0.5)	(0.6, 0.65, 0.7)	(0.8, 0.85, 0.9)	(0.8, 0.85, 0.9)
S_2	(0.8, 0.85, 0.9)	(0.6, 0.65, 0.7)	(0.5, 0.55, 0.6)	(0.4, 0.45, 0.5)	(0.5, 0.55, 0.6)
S_3	(0.7, 0.75, 0.8)	(0.8, 0.85, 0.9)	(0.5, 0.55, 0.6)	(0.7, 0.75, 0.8)	(0.6, 0.65, 0.7)
S_4	(0.8, 0.85, 0.9)	(0.7, 0.75, 0.8)	(0.4, 0.45, 0.5)	(0.5, 0.55, 0.6)	(0.8, 0.85, 0.9)

Therefore, the alternative S_2 is selected as the most appropriate supplier. The net score of the suppliers and their corresponding rankings are also shown in Figures 2.1 and 2.2, respectively.

2.3.2 Illustration of the numerical with the FGT method

In Section 2.3.2, the FGT method is applied to that same cited example. The computation of differences of the suppliers under FGT is shown in Table 2.8. Grey relation coefficient (GRC) of the suppliers is computed and shown in Table 2.9. Grey relational grade (GRG) of the suppliers and their corresponding ranking preferences are shown in Tables 2.10 and 2.11, respectively. According to the downward order of the GRG, the ultimate ranking preference of the alternatives is ascertained. Here, alternative S_2 is adjudged as the best supplier with the highest GRG (9.94). Alternative S_1 is

Table 2.4   Fuzzy weights of the criteria [33].

C_1 (+)	C_2 (+)	C_3 (+)	C_4 (−)	C_5 (−)
(0.83, 0.97, 1)	(0.7, 0.9, 1)	(0.57, 0.77, 0.93)	(0.5, 0.7, 0.9)	(0.7, 0.9, 1)

Table 2.5 Fuzzy weighted decision matrix.

Suppliers	C_1 (+)	C_2 (+)	C_3 (+)	C_4 (−)	C_5 (−)
S_1	(0.50, 0.63, 0.70)	(0.28, 0.41, 0.50)	(0.34, 0.50, 0.65)	(0.40, 0.60, 0.81)	(0.56, 0.77, 0.90)
S_2	(0.66, 0.82, 0.90)	(0.42, 0.59, 0.70)	(0.29, 0.42, 0.56)	(0.20, 0.32, 0.45)	(0.35, 0.50, 0.60)
S_3	(0.58, 0.73, 0.80)	(0.56, 0.77, 0.90)	(0.29, 0.42, 0.56)	(0.35, 0.53, 0.72)	(0.42, 0.59, 0.70)
S_4	(0.66, 0.82, 0.90)	(0.49, 0.68, 0.80)	(0.23, 0.35, 0.47)	(0.25, 0.39, 0.54)	(0.56, 0.77, 0.90)

Table 2.6 Sum of the weighted fuzzy score under FMOORA.

Suppliers	Benefit criteria	Non-benefit criteria	Net fuzzy score
S_1	(1.12, 1.54, 1.85)	(0.96, 1.36, 1.71)	(0.16, 0.18, 0.14)
S_2	(1.37, 1.83, 2.16)	(0.55, 0.81, 1.05)	(0.82, 1.02, 1.11)
S_3	(1.43, 1.92, 2.26)	(0.77, 1.11, 1.42)	(0.66, 0.81, 0.84)
S_4	(1.38, 1.85, 2.17)	(0.81, 1.15, 1.44)	(0.57, 0.70, 0.73)

Table 2.7 Net score and rank of the alternative suppliers under FMOORA.

Suppliers	Net score	Rank
S_1	0.48	4
S_2	2.95	1
S_3	2.30	2
S_4	1.99	3

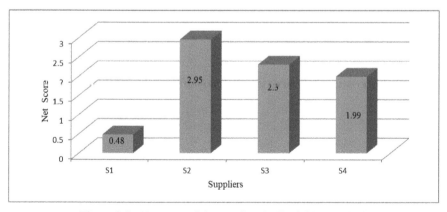

Figure 2.1 Net score of the suppliers by FMOORA method.

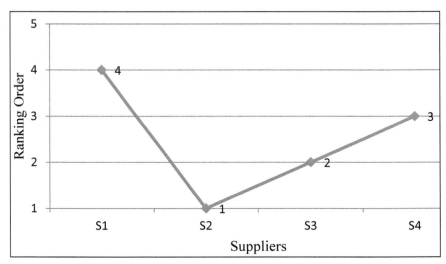

Figure 2.2   The ranking order of the suppliers by the FMOORA method.

Table 2.8   Computation of differences of the supplier alternatives under FGT.

Supplier	C_1 (+)	C_2 (+)	C_3 (+)	C_4 (−)	C_5 (−)
S_1	(0.20, 0.20, 0.20)	(0.40, 0.40, 0.40)	(0, 0, 0)	(0.40, 0.40, 0.40)	(0.30, 0.30, 0.30)
S_2	(0, 0, 0)	(0.20, 0.20, 0.20)	(0.10, 0.10, 0.10)	(0, 0, 0)	(0, 0, 0)
S_3	(0.10, 0.10, 0.10)	(0, 0, 0)	(0.10, 0.10, 0.10)	(0.30, 0.30, 0.30)	(0.10, 0.10, 0.10)
S_4	(0, 0, 0)	(0.10, 0.10, 0.10)	(0.20, 0.20, 0.20)	(0.10, 0.10, 0.10)	(0.30, 0.30, 0.30)
MAX	(0.20, 0.20, 0.20)	(0.40, 0.40, 0.40)	(0.20, 0.20, 0.20)	(0.40, 0.40, 0.40)	(0.30, 0.30, 0.30)
MIN	(0, 0, 0)	(0, 0, 0)	(0, 0, 0)	(0, 0, 0)	(0, 0, 0)

the last choice with the least GRG (5.64). Accordingly, the ultimate ranking preference of the supplier/vendor alternatives is $S_2 > S_4 > S_3 > S_1$, as shown in Table 2.11. Therefore, the alternative S_2 is selected as the most suitable supplier. GRG values of the suppliers and their corresponding rankings under FGT are also shown in Figures 2.3 and 2.4, respectively.

Table 2.9 Grey relation coefficient (GRC) of the supplier alternatives.

Suppliers	C_1 (+)	C_2 (+)	C_3 (+)	C_4 (−)	C_5 (−)
S_1	(0.33, 0.33, 0.33)	(0.33, 0.33, 0.33)	(1.0, 1.0, 1.0)	(0.33, 0.33, 0.33)	(0.33, 0.33, 0.33)
S_2	(1.0, 1.0, 1.0)	(0.50, 0.50, 0.50)	(0.50, 0.50, 0.50)	(1.0, 1.0, 1.0)	(1.0, 1.0, 1.0)
S_3	(0.50, 0.50, 0.50)	(1.0, 1.0, 1.0)	(0.50, 0.50, 0.50)	(0.40, 0.40, 0.40)	(0.60, 0.60, 0.60)
S_4	(1.0, 1.0, 1.0)	(0.67, 0.67, 0.67)	(0.33, 0.33, 0.33)	(0.67, 0.67, 0.67)	(0.33, 0.33, 0.33)

Table 2.10 Grey relation grade (GRG) of the supplier alternatives.

Suppliers	C_1 (+)	C_2 (+)	C_3 (+)	C_4 (−)	C_5 (−)	GRG
S_1	(0.28, 0.32, 0.33)	(0.23, 0.30, 0.33)	(0.57, 0.77, 0.93)	(0.17, 0.23, 0.30)	(0.23, 0.30, 0.33)	5.64
S_2	(0.83, 0.97, 1.0)	(0.35, 0.45, 0.50)	(0.29, 0.39, 0.47)	(0.50, 0.70, 0.90)	(0.70, 0.90, 1.0)	9.94
S_3	(0.42, 0.49, 0.50)	(0.70, 0.90, 1.0)	(0.29, 0.39, 0.47)	(0.20, 0.28, 0.36)	(0.42, 0.54, 0.60)	7.54
S_4	(0.83, 0.97, 1.0)	(0.47, 0.60, 0.67)	(0.19, 0.26, 0.31)	(0.33, 0.47, 0.60)	(0.23, 0.30, 0.33)	7.56

Table 2.11 The ranking order of the suppliers under FGT.

Suppliers	GRG	Rank
S_1	5.64	4
S_2	9.94	1
S_3	7.54	3
S_4	7.56	2

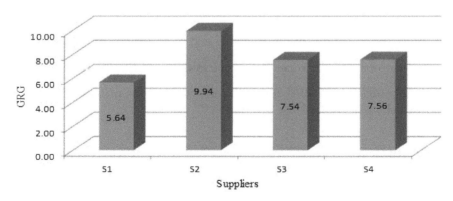

Figure 2.3 Grey relational grade (GRG) of the suppliers.

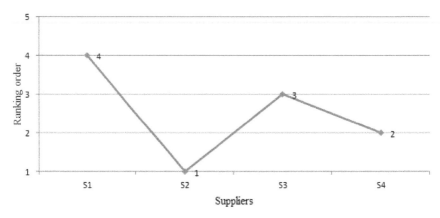

Figure 2.4   The ranking order of the suppliers by the FGT method.

2.4 Overall Discussion and Relative Result Analysis

The illustration shows that the fundamentally strong FMOORA approach got the ranking order of the supplier alternatives as $S_2 > S_3 > S_4 > S_1$ as shown in Table 2.12. For the FGT approach, the preferred order of the alternatives is $S_2 > S_4 > S_3 > S_1$ and also shown in Table 2.12. The ranking of the best and worst alternative suppliers was found exactly the same for both approaches, but the ranking of the intermediate suppliers did not remain the same. Wu and Liu [32] also illustrated the same problem statement using fuzzy technique for order preference by similarity to ideal solution (TOPSIS) with a vague sets approach. Fuzzy TOPSIS with vague sets approach found the ranking preference of the alternatives as $S_2 > S_4 > S_3 > S_1$ as shown in Table 2.12. For all these MCDM methods, the top-most ranking and the least preferred suppliers for selection remain the same. This relative result investigation signifies the applicability and compatibility of those two methods as very practical, capable, and effective decision-making tools under uncertain environments. The relative ranking order of the supplier selection problem using different MCDM approaches is shown in Table 2.12 and also graphically shown in Figure 2.5.

2.5 Conclusion

Supplier selection could well be decided by a fuzzy MCDM approach under vague, ambiguous, and imprecise information. The evaluation criteria of alternative suppliers frequently vary depending upon the classification of

Table 2.12   The relative ranking order of the supplier alternatives under different approaches.

Suppliers	Wu and Liu [32]	Proposed method	
	Fuzzy TOPSIS with vague sets	**FGT**	**FMOORA**
	Ranking order	**Ranking order**	**Ranking order**
S_1	4	4	4
S_2	1	1	1
S_3	3	3	2
S_4	2	2	3

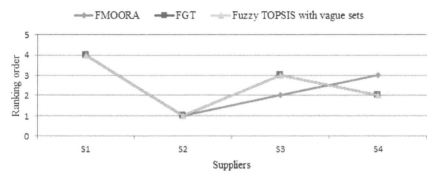

Figure 2.5   The relative ranking order of the suppliers by different MCDM approaches.

products of the chain. The selection process grows to be complex under uncertainty. An expert's knowledge, experience, and verdicts have a significant role in evaluating this information in terms of linguistic expression, which is converted into triangular fuzzy numbers. This information has been assessed by two well-known MCDM techniques, namely FMOORA and FGT comfortably.

While applying the FMOORA approach, it is observed that the method simply could correlate with the given ambiguous environment as it prevails in the numerical illustration. According to its algorithm, the ultimate ranking preference of the vendors for this illustrated example was obtained as $S_2 > S_3 > S_4 > S_1$. Therefore, the alternative S_2 is the most appropriate supplier as it has the highest net score (2.95) value among all the alternatives.

In the case of the FGT approach, it is also observed that the method simply could associate with the given ambiguous environment. FGT approach found the ranking preference of the suppliers as $S_2 > S_4 > S_3 > S_1$. Consequentially, the alternative S_2 is the most suitable supplier as it has the highest GRG (9.94) value among all the supplier alternatives.

2.6 Acknowledgement

I, on behalf of my co-authors, would like to take this opportunity to convey my sincere regards and reverence to all the esteemed professors who authored the research papers, namely Li *et al.* [20], Wu and Liu [32], and Chen *et al.* [33]. The illustrated problem statement for this chapter is extracted partially from their research papers.

All the co-authors and I also thankfully acknowledge the moral support and help provided by our present institution Haldia Institute of Technology (HIT), Haldia, WB, India, in presenting this chapter to the 5th International Conference on Mathematical Techniques in Engineering Applications (ICMTEA 2021) December 3–4, 2021.

References

[1] M. E. Porter, 'Competitive Advantage: Creating and Sustaining Superior Performance,' Collier Macmillan, London, 1985.

[2] S. Chopra and P. Meindl, 'Supply Chain Management. Strategy, Planning & Operation.' Pearson Education, Inc., publishing as Prentice Hall, DOI:10.1007/978-3-8349-9320-5_22, 2002.

[3] A. Sanayei, S. F. Mousavi and A. Yazdankhah, 'Group decision making process for supplier selection with VIKOR under fuzzy environment.'Expert System with Applications, 37, pp. 24–30, 2010.

[4] W. D. Hong, B. Lyes, and X. L. Xie, 'A simulation optimization methodology for supplier selection problem.'International Journal of Computer Integrated Manufacturing, 18 (2–3), pp. 210–224, 2005.

[5] N. O. Ndubisi, M. Jantan, L. C. King and M. S. Ayub, 'Supplier selection and management strategies and manufacturing flexibility.' Journal of Enterprise Information Management 18(3), pp. 330–349, 2005.

[6] R. Lasch and C. G. Janker, 'Supplier selection and controlling using multivariate analysis'. International Journal of Physical Distribution and Logistics Management, 35(6), pp. 409–425, 2005.

[7] R. M. Monczka, R. B. Handfield, L. C. Giunipero and J. L. Patterson, Purchasing and Supply Chain Management, (4th ed.), NY: Centage, 2009.

[8] R. M. Monezka and S. J. Trecha, 'Cost-based supplier performance evaluation.' Journal of Purchasing and Materials Management 24(2) pp. 2–7, 1998.

[9] C. L. Hwang and K. Yoon, 'Multiple attribute decision making: Methods and application.' New York: Springer, 1981.

[10] F. Herrera and E. Herrera- Videna, 'Direct approach processes in group decision making using linguistic OWA operators.' Fuzzy sets Syst., 79(2), pp. 175–190, 1996.

[11] L. De Boer, L. van der Wegen and J. Telgen, 'Outranking methods in support of supplier selection.' European Journal of Purchasing and Supply Management, 4, pp. 109–118. 1998.

[12] F. Herrera, Herrera-Viedma and J. L. Verdegay, 'A model of consensus in group decision making under linguistic assessments.' Fuzzy Sets and Systems, 78, pp. 73–87, 1996.

[13] L. A. Zadeh, 'Fuzzy sets'. Information and Control, 8, pp. 338–353, 1965.

[14] T. C. Chen, 'Extensions of the TOPSIS for group decision-making under fuzzy environment'. Fuzzy Sets and Systems, 114, pp. 1–9, 2000.

[15] C. Kahraman, D. Ruan and D. G. Ibrahim 'Fuzzy group decision-making for facility location selection.' Information Sciences, 157, pp. 135–153, 2003.

[16] H. J. Shyur and H. S. Shih, 'A hybrid MCDM model for strategic vendor selection.' Mathematical and Computer Modelling, 44(8), pp. 749–761, 2006.

[17] L. De Boer, E. Labro and P. Morlacchi, 'A review of methods supporting supplier selection.' European Journal of Purchasing and Supply Management, 7(2), pp. 75–89, 2001.

[18] S. Talluri and R. Narasimhan, 'A methodology for strategic sourcing.' European Journal of Operational Research, 154(1), pp. 236–250, 2005.

[19] Z. Huixia and Y. Tao, 'Supplier Selection Model based on the Grey System Theory.' The 2008 International Conference on Risk Management & Engineering Management, 978-0-7695-3402-2/08 © 2008 IEEE ; DOI 10.1109/ICRMEM.2008.13, 2008.

[20] G. Li, D. Yamaguchi and M. Nagai, 'A grey based decision-making approach to the supplier selection problem.' Mathematical and computer modeling, 46 (3-4), pp. 573–581, 2007.

[21] W. K. M. Brauers and E. K. Zavadskas, 'The MOORA method and its application to privatization in a transition economy.' Control and Cybernetics, 35, pp. 445–469, 2006.

[22] B. Dey, B. Bairagi, B. Sarkar and S. Sanyal. 'A MOORA based fuzzy multi-criteria decision-making approach for supply chain strategy selection.' Industrial Journal of Industrial Engineering Computations, 3, pp. 649–662, 2012.

[23] S. Chakraborty, 'Applications of the MOORA method for decision making in manufacturing environment.' International Journal of Advanced Manufacturing Technology, 54(9-12), pp. 1155–1166, 2011.

[24] B. Bairagi, B. Dey, B. Sarkar and Sanyal, 'A De Novo multi-approach Multi-criteria decision-making techniques with an application in performance evaluation in material handling device.' Computers and Industrial Engineering, 87, pp. 267–282, 2015.

[25] B. Dey, B. Bairagi, B. Sarkar and S. Sanyal, 'Multi-objective Performance Analysis: A Novel Multi-criteria decision-making approach for a supply chain.' Computers and Industrial Engineering, 94, pp. 105–124, 2016.

[26] B. Bairagi, Sarkar, and Sanyal, 'Group heterogeneity in multi-member decision-making model with an application to warehouse location selection in a supply chain.' Computers and Industrial Engineering, 105, pp. 101–122, 2017.

[27] T. Ozcan, N. Celebi and S. Esnaf, 'Comparative analysis of multi-criteria decision-making methodologies and implementation of a warehouse location selection problem.' Expert Systems with Applications, 38, pp. 9773–9779, 2011.

[28] W. K. M. Brauers, 'Optimization methods for a stakeholder society, A revolution in economic thinking by multi-objective optimization.' Boston: Kluer Academic Publishers, 2004.

[29] J. L. Deng, 'The introduction of grey system.' The Journal of Grey System, 1(1), pp. 1–24, 1989.

[30] R. E. Bellman and L. A. Zadeh, 'Decision-making in a fuzzy environment.' Management Science, 17(4), pp. 141–164, 1970.

[31] Y-J. Wang, 'Combining grey relation analysis with FMCGDM to evaluate the financial performance of Taiwan container lines.' Expert Systems with Applications, 36, pp. 2424–2432, 2009.

[32] M. Wu and Z. Liu, 'The supplier selection application based on two methods: VIKOR algorithm with entropy method and Fuzzy TOPSIS with vague sets method.' International Journal of Management Science and Engineering Management, 6(2), pp. 110–116, 2011.

[33] C. T. Chen, 'A fuzzy approach to select the location of the distribution center.' Fuzzy Sets and Systems, 118, pp. 65–73, 2001.

3

The Use of Computational Intelligence in Process Management

C. Dumitrescu

Splaiul Independentei, Bucharest, Romania

Abstract

Integrated circuits based on fuzzy logic are used in expert systems in the fields of command and control for real-time operations. The soft computing concept was developed to exploit tolerance for inaccuracy, uncertainty, and partial truth in order to gain flexibility, robustness, low cost of solutions, and a better connection with reality. This feature makes it different from conventional (hard) computing, which is characterized by a lack of inaccuracy and partial truths. The ultimate goal would be to match or even surpass the performance of the human mind. Soft computing is characterized by a partnership of several domains, the most important of which are neural networks, genetic algorithms, fuzzy logic, and probabilistic reasoning. Based on the human thinking model, soft computing brings these areas together in a complementary rather than competitive relationship, in which each partner contributes its own advantages and techniques to solve problems that are impossible to solve in another mode. Located between artificial intelligence systems and conventional computing, soft computing is the problem to be solved in such a way that the current state of the system can be measured and compared with the state of the needle to be obtained. The state of the system is the basis for adapting the parameters, which converge with each step toward the optimal solution.

Keywords: complex performance, distributed control system, Fuzzy logic, intelligent control, PLC.

3.1 Introduction

In recent years, the number and variation of applications of fuzzy logic have undergone a process of wide expansion. The range of applications starts from widely used products such as video cameras, video recorders, washing machines, and microwave ovens until set industrial processes, medical tools, and decision-making systems.

Why use fuzzy adjustment? One of the reasons for the industry's orientation toward this area is due to competition between companies (mostly Japanese) that use or have begun to use fuzzy adjustment in competitive products; the advantages of its use are found in the specialized economic reports [1–4]. Anyway, the discussions in literature and specialized forums have led to the following reasons:

- Fuzzy adjustment is a new technology, and, consequently, the claims disappear related to the patent of the solution for similar technical problems.
- In Japan, fuzzy regulation is "demanded" by consumers, since it is a "high-tech" solution. In this case, the techniques are fuzzy and used for market reasons (in response to market demand).
- The development of fuzzy regulators is easier to learn and requires less trained staff than that in the case of conventional regulators. As a result, production is cheaper.
- Fuzzy controllers are more robust than conventional ones.
- Fuzzy regulators are more suitable for regulating processes nonlinear.

Scientifically, only the last three reasons are of interest and the following discussion will be focused on this regard.

Fuzzy controllers are represented by if−then and so on rules that provide a user-friendly representation of knowledge. Representation can be seen as a very high-level programming language, where the program consists of if−then rules, and the compiler and/or interpreter uses a nonlinear adjustment algorithm. Thus, programming with the help of expressions is qualitative, represented by if−then rules, and a work program with expressions is obtained quantitative, provided by transducers and execution elements [5–7]. This process will lead to information loss, because there is no single translation from a qualitative entity to a quantitative representation, except in special cases. For example, there is no unique representation for the qualitative expression "high voltage" in a real voltage and vice versa. Because in regulation the output signal from the regulator must be precise from a quantitative point of view (since, physically, it represents a signal of actuation of an electric motor, hydraulic motor, pump, etc.), additional special techniques are required for translation from the qualitative to the quantitative one [8–10]. The advantage of fuzzy tuning is the ability to "program" knowledge of process operators and engineers (consisting of process operation experience) in an easy-to-understand and friendly way [11, 12].

It is often said that fuzzy adjustment is more robust. However, they were not found such research results show that fuzzy regulators are more robust than conventional regulators in general. A fuzzy regulator is a static nonlinearity and whether it is more robust than a conventional regulator depends on the rules that define this static nonlinearity. So "fuzzy regulators are more robust" should be interpreted as "fuzzy controllers are more robust to changes in known parameters." But how to build a more robust fuzzy controller remains a problem since robustness depends on the degree of knowledge of the process regulated. One reason for using fuzzy adjustment is that it is more suitable for the regulation of nonlinear processes. A fuzzy regulator or generally a nonlinear one is, in principle, being capable of regulating a nonlinear process is a problem that depends on the chosen inputs of the regulator. Fuzzy regulators are said to be superior to regulators conventional in the regulation of nonlinear processes [13–15]. This is true to the extent that there is additional knowledge about the nonlinearities of the process.

Most implementations of distributed fuzzy systems are based on the interaction of distributed modules, which have a set of established rules and communicate with each other. The most classic solution for coding a fuzzy system is to use a high-level language. Another solution is to use dedicated hardware such as fuzzy processors. None of the solutions provide an easy way to implement fuzzy systems: the first because it does not have methods for distributing applications to multiple processors, and the second because hardware solutions require quite a long time for development. The problem is to define distributable fuzzy components, such as fuzzy sensors or inference fuzzy components, which are connected to a communication bus. Depending on the configuration, a fuzzy component can implement one or more combinations of fuzzy processes, known as real-fuzzy transformation (TRF), inference, and fuzzy-real transformation (TFR).

3.2 Proposed Solution

Components of a Fuzzy Intelligent Control System:

A membership function represents each fuzzy set separately and acts as a transformation between the rigidly described real world and the fuzzy approach to this world. In control algorithms based on fuzzy logic, membership degrees are used as inputs. Determining the degree of membership functions appropriate to the algorithms is part of the design process.

Designing rules-based systems are currently the most popular application of fuzzy logic. Its basic structure, presented in Figure 3.1, contains three basic sections:

1. section for "rigid-fuzzy" transformation;
2. section of the inference mechanism, which uses the system of rules;
3. section for the "fuzzy-rigid" transformation.

Using such a system, we transform the rigid domain of the inputs into a fuzzy domain, process the information thus obtained, and transform the results of the processing, which belong to a fuzzy domain, into a rigid domain of the outputs. This approach is similar to how it works in the frequency domain using time-transformed data. Because data processing is easier in the frequency domain than in the time domain, time—frequency and frequency—time transformations are justified. In a fuzzy system, working with a rule base that describes how the system works in fuzzy terms is easy to do. Consequently, it is convenient to transform rigid inputs and outputs into a fuzzy field of intuitive linguistic rules. Rigid-fuzzy transformation is an application from rigid inputs to a degree of belonging through the functions of degree of belonging. The resulting degree of affiliation thus becomes the input for the section of the system in which the inference mechanism is applied. In the inference module, the input values and also the values established by the

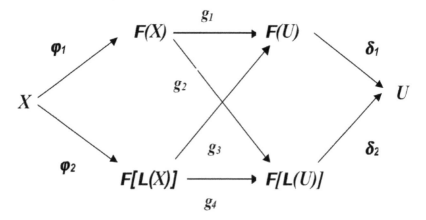

Figure 3.1   Fuzzy processors classification.

fuzzy rules are analyzed as conditions related to the rules in the composition of the rule base. At certain preset intervals, the fuzzy processor reads the input data and applies the set of established rules, thus obtaining the desired outputs. Theoretically, the rule base predicts all possible combinations of states that can be obtained from the analysis of input values, but most of the time, an extended rule base is not necessary.

3.2.1 Methods of choosing solutions for determination output

Because, for a lot of entries, several rules can be met at the same time, methods had to be found to combine the resulting actions into one. In addition, the resulting action must be transformed from a fuzzy size into a rigid, executable size. In practice, situations may occur in which any of the following four known methods can be used:

- The maximum method takes into account the highest value of the degrees of belonging of the various rules that are fulfilled at the same time, which it uses as a resultant action. As a possibility to resolve the conflict, the rigid exit can take a value corresponding to the average of the individual exits corresponding to the two equal degrees of belonging. Another possibility is to select the output associated with the weight or position of the rule within the rule base. The maximum method is the simplest approach for determining the outputs, but it is not agreed upon due to possible deviations from an optimal result.
- The weighted average method applies weights to the various possible actions depending on their degree of affiliation and calculates their arithmetic mean. Like the maximum method, this method suffers from ambiguity because an output membership function can specify more than one value for a given value of μ. An output membership function has a graphical representation in the form of a pyramid or pyramid trunk. If μ has an average value, the value of the output is derived from the ascending slope or descending slope of the function graph. In addition, for a pyramid trunk, $\mu = 1$ corresponds to an entire range of values.
- The method of the center of gravity calculates the action output by associating with the center of gravity of the surface delimited by the graphs of the output membership functions, limited superior by the values of the membership degrees for which the rules have been fulfilled. This method consumes higher computing resources than those presented above and also suffers from certain limitations. For membership functions with vertical symmetry, the center of weight will have the same value, regardless

of the value of degree of affiliation resulting from the application of the rules (μ). Therefore, to ensure an output across the definition range without sudden variations, at least two rules from the rule base must be met for each sampling of the inputs. Due to limitations, the center of gravity method is considered the optimal solution for combining and transforming fuzzy-rigid.

- The singleton method is a particular case of the previous method. This method succeeds in interpreting fuzzy output data as unique values, using the weighted average results for all possible output combinations. This interpretation consumes fewer computing resources than the center of gravity method, but at the same time also requires overlapping membership functions to avoid discontinuities at the exit. Due to its conceptual and combinational simplicity, this method tends to replace the center of gravity method in many cases.

3.2.2 Fuzzy cells

Today most control algorithms are implemented on the computer. This involves measuring the inputs to the controller at a certain sampling period. For example, the classic linear fuzzy adjustment algorithm can be represented in Figure 3.2. Thus, the regulator outputs are considered static functions (mappings) of the regulator inputs. The dynamic behavior of the regulator (such as derivative or integral action) is emulated by extending the function of the regulator to several inputs. These inputs are delays or differences in some inputs and outputs. In this way, a regulator is considered to be composed of a static function of the regulator and additional parts such as pre-filtering or post-filtering to obtain delayed signals, difference type inputs, integrations, and signal limitations. Regarding the stability of fuzzy systems, it should be noted that fuzzy regulators can be considered nonlinear regulators and for this reason, it is difficult to find a general result in the analysis and synthesis of fuzzy regulators. Demonstrations of stability for fuzzy controllers found in the literature are restricted to those relatively simple fuzzy controllers (e.g., PID fuzzy controllers) and where the adjustment process is stable. If the process cannot be mathematically modeled, then no demonstrations of stability can be given. The development of the modern theory of regulation owes a lot to mathematical modeling, but the implementation in real life of regulation solutions often encounters difficulties due to the vague, imprecise nature of the regulated process. Fuzzy sets and fuzzy concepts in general arose from the need to quantitatively express "vague" and "imprecise."

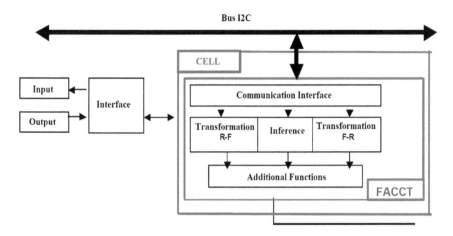

Figure 3.2   Fuzzy cell architecture.

Although there are many branches of mathematics older than fuzzy set theory, which deal with the study of random nature, probability theory, mathematical statistics, information theory, and others, no substitutions can be made between them and the fuzzy set theory. The fuzzy cell configuration is shown in Figure 3.3.

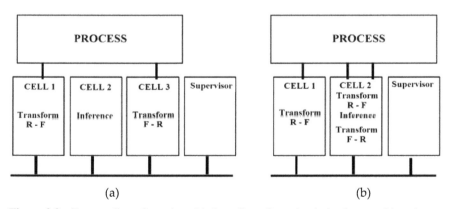

Figure 3.3   Fuzzy cell configuration: (a) the cell configuration is implemented based on a fuzzy sensor and actuator, or we can introduce the inference component; (b) the implementation configuration of two fuzzy cells — cell 1 has a fuzzy sensor, and cell 2 has a fuzzy controller.

3.2.3 Mathematical model for fuzzy components

Real-fuzzy symbolic transformation:

Fuzzy linguistic variables are fuzzy sizes associated with deterministic ones. The equivalent of the scalar value in the deterministic sense is for a fuzzy variable, the linguistic degree (label and attribute) associated with it. Thus, as for bivalent logic, the deterministic value "1" is associated with the TRUE attribute, and "0" is labeled FALSE; in fuzzy logic, for the deterministic variable positive real number, the associated linguistic variable can be, for example, the distance between two points, which may have language degrees NEAR, MEDIUM (MEAN), FAR or TOO NEAR, NEAR, MEDIUM, FAR, TOO FAR. The domain of values of the deterministic quantity is called the discourse universe. Each attribute of a linguistic variable is associated with a function of belonging, whose value (in a deterministic sense) indicates the level of confidence with which a deterministic value can be associated with that attribute of the linguistic variable. For example, considering the linguistic variable distance of three linguistic degrees, NEAR, MEDIUM, and FAR, we associate with them typical membership functions, as shown in Figure 3.4.

For example, the language modifier is used to change its meaning from large to very large. Some authors have called these linguistic modifiers "hedges." There are two approaches in this sense: powered fence and shifted fence, the scaled fence being a combination of the advantages of the two approaches.

Linguistic modifiers operate on the degrees of belonging and are represented by

$$\forall x \in X, \mu_{M(L)}(x) = \mu_R(x, L). \tag{3.1}$$

The membership function $\mu_{M(L)}(x)$ shows that an element of X can be associated with a linguistic term. The advantage of these graduated modifiers is that for every modifier, a basic operation will be defined by setting a standard value of p. The properties $m_p(A)$ can be deduced for different values of p:

$$0 < p < 1; \text{ the fuzzy set expands } m_p \supset A \tag{3.2}$$

$$p = 1; \text{ the fuzzy set does not change } m_p \equiv A \tag{3.3}$$

$$p > 1; \text{ the fuzzy set is concentrated } m_p \subset A. \tag{3.4}$$

The support and core of a fuzzy set do not change by applying graduated modifiers. For example, the intensification ("really") has the effect of increasing the degrees of belonging above 0.5 and decreasing thosebelow 0.5:

$$\mu_{\text{int }(A)} = \begin{cases} 2\left(\mu_A(x)\right)^2, \mu_A(x) \le 0.5 \\ 1 - 2\left(1 - \mu_A(x)\right)^2, \mu_A(x) > 0.5 \end{cases}. \tag{3.5}$$

According to Figures 3.2 and 3.4, the membership function $\mu_{D(x)}(L)$ shows the linguistic term L of an element in X.

It follows that the meaning of fuzzy and the description of fuzzy are linked by an equal relation (see Figure 3.4):

$$\mu_{D(x)}(L) = \mu_{M(L)}(x). \tag{3.6}$$

The notation $M(L)$ is replaced by the notation L. Thus, instead of writing $\mu_{M(L)}(x)$, we write $\mu_L(x)$. In the case of the very modifier, the method consists in creating support for $M(L)$, the size of the nucleus of L, and reducing the nucleus by the same amount if possible. For triangular membership functions, this reduction of the core is not possible. For the linguistic modifier, more or less (somewhat) a complementary method is applied (since the linguistic modifier is considered complementary to very).

The $\mu_L(x)$ values are also called support values, the degree of fulfillment of the rule or the degree of fit between the data and the premises of the rule. The inference is thus reduced to a simple calculation scheme. The next section shows the "max$-$min" inference method.

Symbolic fuzzy inference:

A fuzzy system is generally defined as a functional relationship drawn between two unitary multi-dimensional spaces:

$$S : I^n \longrightarrow I^m, \tag{3.7}$$

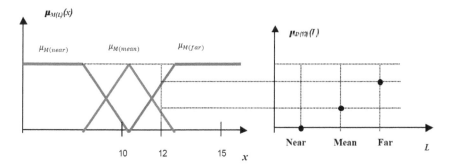

Figure 3.4 The fuzzy significance and the fuzzy class for $x = 13$.

wherein the unitary n-dimensional space I^n comprises all the n fuzzy subsets of the input variables, and the p-dimensional space contains all the m fuzzy subsets of the output variables. This association between fuzzy set spaces is enshrined as fuzzy associative memories (FAMs), but in this paper, we will use the name fuzzy rule bases (BRFs). Practically, any system can be considered as an input–output relationship whose response is presented as a generic function *output = f(inputs)*. This function is the foundation of the basic model of systems, which generally represents a mathematical expression of their way of working. A rules-based system can approximate any continuous function with as many variables as accurately possible. The consistency of a BRF can be deduced based on the data regarding the operating points of the system. These data are accessible by monitoring the existing system in operation, of a similar one or based on an analytical model. Fuzzy systems have a close interdependence between the membership functions of the variables in the system and BRF. Thus, a BRF is constructed by logically linking the fuzzy sets associated with the output variables with the fuzzy sets of the input variables. For this purpose, we start from the general strategy for describing the process, which eventually breaks down into sub-strategies specific to certain control stages. Heuristic methods and knowledge engineering techniques play an important role here. The control strategy is expressed in linguistic terms based on which logical inferences are formed, which will constitute the BRF (logical associations) rules. The inference is the logical operation that allows the transition from premise to conclusion based on formal reasoning. A rule occurs when there is a premise about an event, which involves or attracts a certain logical consequence (conclusion). In general, any physical process can be modeled on its description by rules. This involves establishing a set of premises and identifying the set of consequences (practically, a set of cause–effect relationships is identified). Therefore, a rule is expressed by composing the premises $P_i(i = \overline{1, ..., n})$ with the help of established logical operators (generically denoted by the symbol $\otimes$) and equating the result with the consequence C:

$$P_1 \otimes P_2 \otimes ... \otimes P_n = C \qquad (3.8)$$

In our cases, we will consider the "max–min" inference method, which is often used in fuzzy adjustment. Fuzzy relationships can be combined through the composition operation. Several ways of composing fuzzy relationships have been proposed, the most used being the max–min composition. The fuzzy regulator uses the so-called max–min method.

Let R_1 and R_2 be two binary fuzzy relations defined on $X \times Y$ and $Y \times Z$. The max−min of the relations R_1 and R_2 is a fuzzy set defined as follows:

$$R_1 {}^{\circ} R_2 = \{((x, z) \, \mathrm{maxmin} \, \{\mu_{R_1}(x, y), \mu_{R_2}(y, z)\}) \, ; x \in X, y \in Y, z \in Z\}, \tag{3.9}$$

whose membership function is

$$\mu_{R_1 {}^{\circ} R_2}(x, z) = \mathrm{maxmin} \, \{\mu_{R_1}(x, y), \mu_{R_2}(y, z)\}. \tag{3.10}$$

The conjunction, the implication, and the composition must be based on the same T norm to obtain a simple analytical solution following the fuzzy inference. The result of the inference results from the law of composition:

$$\forall L'' \in L(U), \mu_F\left(L''\right) = \max \min \left\{ \min \left[\mu_{\rho 2(\varepsilon)}(L), \mu_{\rho 2(\Delta\epsilon)}\left(L'\right)\right], \right.$$
$$\left. \mu_{\Gamma}\left(L, L', L''\right)\right\} \tag{3.11}$$

$$L \in L_{(\epsilon)}, L' \in (\Delta\varepsilon).$$

Fuzzy-real symbolic transformation:

They are used to determine that part of the fuzzy output whose membership values are below a certain threshold. Apart from the center of gravity method, another basic method of defuzzification is the mean of maxima (MoM), represented by δ_1 in Figure 3.1. The method of defusing the center of the surface is defined by

$$\forall u \in U, \mu_G(u) = \mu_{\delta 3(F)}(\mathrm{u}) = \max_{L \in L(U)} \min \left[\mu_F(L), \mu_{M(L)}(u)\right] \tag{3.12}$$

Using the weight method, the values returned from the fuzzy actuator are

$$U = \delta_2(F) = \frac{\int \mu_G(\gamma) \cdot \gamma \cdot d\gamma}{\int \mu_G(\gamma) \cdot d\gamma}, \gamma \in U_s. \tag{3.13}$$

Let the fuzzy subset be $F = \{\frac{0}{N}, \frac{0.4}{Z}, \frac{0.6}{P}\}$. Figure 3.5 proposed a defuzzification solution using a triangular min norm and the result $\mu = 0.12$.

Fuzzy cell configuration:

In the theoretical approach, it was shown that the fuzzy regulation algorithm consists of three phases: fusing, the composition between the fuzzy input relation and the fuzzy relation of the regulator, and the defuzzification using language FACCT (fuzzy applied cell control technology). This involves

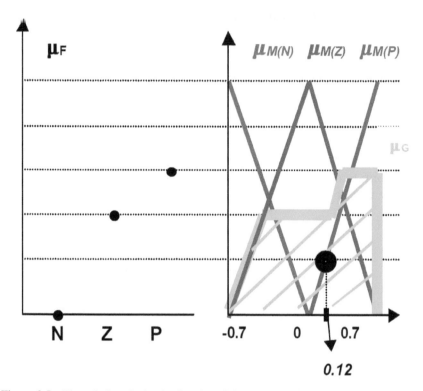

Figure 3.5   The solution obtained using the minimum triangular norm leads and the result $\mu = 0.12$.

a great deal of computational effort in the case of multi-dimensional functions, which is not recommended in practice. In practice, fuzzy regulation
involves the use of local inferences. The following entries will be considered
numerical (singletons) for simplicity. This simplification is not a severe
restriction since the inputs to the controller are usually numeric values taken
from the sensors. Figure 3.6 shows the fuzzy configuration.

In practice, the inference of a rule base is local. Thus, the inference of
rule bases implies the inference of each rule followed by the aggregation of
the obtained results. For numerical inputs, the result of global inference is
identical to that of local inference. In those, the following will be presented
a practical fuzzy inference scheme as well as an analysis of the various
fuzzy implications used in fuzzy regulation. Given these aspects, FACCT is
considered a programming language for the development of fuzzy distributed

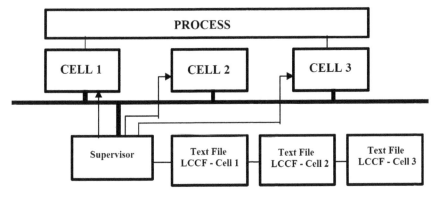

Figure 3.6 Architecture supervisor fuzzy configuration.

intelligence systems (Figure 3.7). The fuzzy cells together with the resources allocated to the processing can be constituted in a resident compiler.

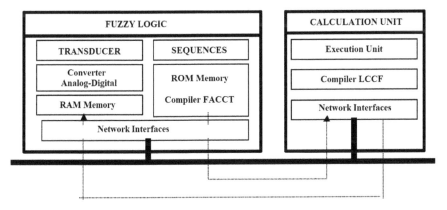

Figure 3.7 The architecture of a fuzzy distributed intelligence system that is programmed in the FACCT language.

3.2.4 Elements of programming language FACCT

Fuzzy logic is seen as one of the techniques of artificial intelligence that includes "conventional" expert systems, neural networks, and genetic algorithms. Zadeh (1994) proposes the name of soft computing for the field of neural networks, genetic algorithms, fuzzy logic, and their combinations.

Today, the field of regulation and fuzzy modeling is considered in many publications to overlap with that of neural networks. There are many publications in the literature about neuro-fuzzy systems or neuro-fuzzy networks. An important concept in fuzzy set theory and fuzzy logic is the linguistic variable (Zadeh, 1994). The connection with artificial intelligence also appears from the statement of Dubois (1991) who says that "the main reason that determined the theory of fuzzy sets is the desire to build a formal, quantitative framework that captures the vague, the imprecise in human knowledge as they are expressed through natural language." From this point of view, fuzzy logic and rough reasoning can provide a framework for understanding and processing natural language and for modeling the way people reason and communicate. Zadeh (1994) highlights the difference between fuzzy regulation and knowledge-based systems as follows: "What makes the difference between regulation applications and those of knowledge-based systems is that in main regulation the problem that is resolved is the inaccuracy. On the contrary, in the case of knowledge-based systems, the problems are related to both inaccuracy and the unknowns of the model." Fuzzy regulation can be considered a small part of the theoretical framework of approximate reasoning.

Figure 3.8 shows a subroutine software example of proportional fuzzy control.

Below we will mention some FACCT elements:

1. **Declaration-type variables:**
   Crisp [numeric variable] = Defines numeric variables belonging to rigid input and output sets, as well as other numeric variables useful in the system.
   Subset [...] = Defines variables as fuzzy subsets.
   Varlin [linguistic variable] = Defines the linguistic variables that will receive fuzzy partitions.
   The term [linguistic term] = Defines the list of linguistic terms used to describe linguistic variables.

2. **Initialization-type variables:**
   Partition [varlin, term, inf, sup, a, b, c, d] = Defines the meaning of the linguistic term associated with the linguistic variable varlin, in the universe of discourse [inf, sup].
   varlin_1 = varlin_2 = Assigns the meanings of the linguistic terms of the variable varlin_1 to the linguistic variable varlin_2. Both variables must be previously defined with the varlin function.

```
1.      declarations;
2.      crisp x, u;
3.      subset f,g;
4.      varlin Error, Control;
5.      term Z, P, N: Error;

6.      initialisation ;
7.      partition(Error,"Z", -1, 1, -1, 0, 0, 1);
8.      partition(Error, "P", -1, 1, 0, 1, 1,);
9.      partition(Error, "N", -1, 1, , -1, -1, 0);
10.     Control=Error;
11.     f as Error;
12.     g as Action;
13.     main ;

14.     x=input(0);
15.     f=fuzz(x);
16.     g=0;
17.     if f is "N" then g is "N";
18.     if f is "Z" then g is "Z" (0.75);
19.     if f is "P" then g is "P"
20.     u = defuzz(g);
21.     output(u);
```

Figure 3.8   A subroutine software example of a FACCT program.

subfuse as varlin = The fuzzy subfuse subset is defined relative to the set of linguistic terms of the varlin linguistic variable.

3. **Execution variables:**

crisp_1 = *crisp_2* = Assigns to numeric variable crisp_1 value of numeric variable crisp_2.

crisp = *input (i)* = The numeric variable crisp is assigned the numeric value converted to the analog and processor input.

fuzz (crisp) = The fuzz operator merges the crisp numeric variable defined with the crisp function.

defuzz (crisp) = The defuzz operator defuses the crisp numeric variable defined with the crisp function.

output (crisp) = The defuzzification result is transmitted to the analog output associated with the crisp numeric variable defined with the crisp function.

if subfuz_1 is term_1 and subfuz_2 is term_2 and... then subfuz_k is term k [(height)] = Defines inference rules of type if−then, where *subfuz_i* are the fuzzy subsets defined with the subset function, and

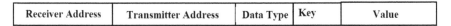

Receiver Address	Transmitter Address	Data Type	Key	Value

Figure 3.9   Transmitted data format.

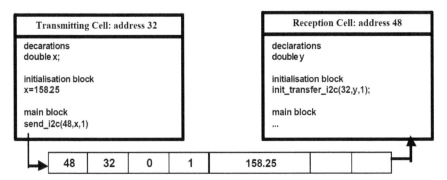

Figure 3.10   Description of a programming example in FACCT.

term_i are the linguistic terms associated with the fuzzy subset defined relatively to a linguistic variable.

recv (adr, name, key) = Initiates the reception of the contents of the *name* variable from the cell with the address *adr*.

send (adr, name, key) = Initiates the issuance of the contents of the *name* variable to the cell with the address *adr*, associating the key with it.

Figures 3.9 and 3.10 show how data is formed and an example of data exchange between two fuzzy cells.

3.2.5 Results

Application of dynamic fuzzy controller for a greenhouse using Mamdani processor:

The purpose of this application is to create a general-purpose regulator that can operate in a wide range of applications, exemplified for the control of a greenhouse. For the implementation, we will make a fuzzy controller in the Mamdani version. Mamdani's method has allowed the development of many engineering applications, especially in the field of fuzzy command and control systems. Mamdani introduced a fusion/inference/defuzzification scheme and used an inference strategy, which is known as the max−min method. This type of inference represents a method that connects the input language variables with the output language variables using only MAX- and MIN-type functions.

Let be a set of inference rules in the form of a fuzzy set represented as a matrix or a table, and a few fuzzy sets together with their associated membership functions, assigned to each of the variables in the fusing process (real-fuzzy transformation). Figure 3.11 shows the case where two rules are active, which involve two input variables (*x* and *y*) and an output variable (*r*). Suppose that *x* is a measure of *x(t)* at time *t* and *y* is a measure of *y(t)* at the same moment of time. Fuzzy sets *A1*, *A2*, *B0*, and *B1* have the respective membership functions $\mu_{A1}(x)$, $\mu_{A2}(x)$, $\mu_{B0}(x)$ and $\mu_{B1}(x)$. Activated inference rules are, for example:

IF (*X* = *A1* AND *Y* = *B0*) THEN *r* = C0, ELSE
IF (*X* = *A2* AND *Y* = *B1*) THEN *r* = C1.

These rules can also be expressed as follows:

$$\mu_{A1}(x) \wedge \mu_{B0}(x) \rightarrow \mu_{C0}(x, y) \tag{3.14}$$

$$\mu_{A2}(x) \wedge \mu_{B1}(x) \rightarrow \mu_{C1}(x, y) \tag{3.15}$$

A statement like *x* = *A1* is true with a degree of belonging $\mu_{A1}(x)$ and a rule is activated when the combination of all degrees of belonging $\mu_i(k)$ in the premise part of the rule takes a strictly positive value. Several rules can be activated simultaneously. The max−min method performs the AND operators from the different ones' conditions of a rule by taking into account the membership functions with maximum values. The implications (THEN with the role of connectivity) are realized by truncating (limiting) the sets output. This consists in taking into account at each point the minimum value of the degrees of affiliation resulting from the conditions of the rules (in Figure 3.11, $\mu_{A2}(x)$ and $\mu_{B0}(y)$) and the affiliation functions of the respective fuzzy output sets (in Figure 3.11, $\mu_{C1}(r)$ and $\mu_{C0}(r)$):

$$\mu'_{C0}(r) = \mu_{C0}(r) \wedge [\mu_{A1}(x) \wedge \mu_{B0}(y)] \tag{3.16}$$

$$\mu'_{C,1}(r) = \mu_{C1}(r) \wedge [\mu_{A2}(x) \wedge \mu_{B1}(y)] . \tag{3.17}$$

The rules are finally combined by using a connective ELSE operator, which functions as a conjunction operator and which is interpreted as a maximum operation for each possible value of the output value (*r* in Figure 3.11), considering the fuzzy sets defined (*n* sets):

$$\mu_C(r) = \Lambda_r \left[\mu'_{Ci}(r)\right]_{j=0,1...n} \tag{3.18}$$

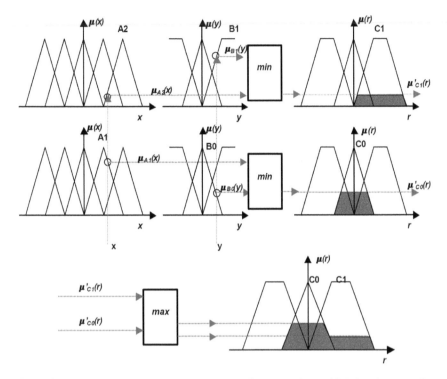

Figure 3.11 Controller Mamdani where two rules are active, which involves two input variables.

Thus, based on the previously defined relationships, it is possible to generate an algorithm for performing fuzzy reasoning to fulfill the control of a process. The max−min method ultimately requires a defuzzification component, which is generally performed using the center of gravity method. For the example in Figure 3.11, the final value will result:

$$V_\Gamma \left[\mu'_{C0(r)}, \mu'_{C1(r)} \right] . \tag{3.19}$$

Properties:

MIN and MAX operators have several properties that they recommend for use in fuzzy inference operations. However, there are other methods that perform OR and AND operators in different ways, and which aim to optimize the way of their implementation in practice. For example, the max-prod method is similar to the max−min method, with the difference that all the implications within the rules are made using the product instead of the minimization operation. The truth values under the rules are used for uniform multiplication of the corresponding output sets instead of limiting them to a certain level. This allows maintaining the information contained in the form of these sets, which is partially lost with the max−min method. From the point of view of software implementation, the product is simpler and faster to execute than the MIN operator and allows the simplification of numerical inferences (how long can you consider analytical expressions instead of comparing each pair of memorized points).

Another method is called sum-prod, which uses arithmetic methods and the product to perform the OR and AND operators. Unlike the MAX operator, which selects only the maximum values, the sum operator considers all the sets involved and preserves some of the information contained in their form.

Their analysis shows that all these methods lead to very close input/output characteristics in the case of a system with a single input and in which we are dealing with trapezoidal or triangular output mutations. In the case of several input variables, the max−min method can produce more pronounced discontinuities than the sum-prod method.

It is no less true that the use of one or the other of the methods is strongly influenced by the way it is implemented. That is why the max−min method is recommended for hardware implementations, due to the ease of implementation of MAX and MIN operators. These operators are more stable if the degrees of affiliation take noisy and imprecise values in the case of hardware with questionable accuracy. The max−min method is also suitable for systems based on fuzzy rules where there is a predetermined model, leading, in most practical applications, to the easy definition of consistent rules.

This is like the behavior of a discrete PD controller whose control is calculated based on the error (the difference between the reference and the process output) and its speed of variation. Such a fuzzy controller can be supplemented with other variables compared to a classic PD regulator. For example, the controller can be improved if the variation of the process output will be used in the calculation of the command instead of the variation of the error or if the reference signal will be classified to capture the nonlinear characteristics of the process. Unlike a classic PID controller, a fuzzy controller can work according to both control principles.

The min operator for conjunction and implication, the max operator for aggregation, and the max−min composition are normally used. This description corresponds to the max−min inference method (Figure 3.11).

Mamdani and his collaborators used this type of fuzzy rule and were the first to report an application of fuzzy logic in regulation (hence the name Mamdani rule). Mamdani fuzzy rules have consequences like fuzzy sentences, and the implication is a conjunction (T-type implication). Using a limited number of membership functions for the fuzzy controller output will lead to restrictions on the control hypersurface, especially if the inputs are normal and form a fuzzy partition and if the inputs are numeric. If a fuzzy controller with a single input and a single output is considered, the function of the controller is limited to an interpolation of the characteristic points (Figure 3.11). This limitation is caused by using a single language tag defined for the controller output when applying the fuzzy rule.

Areas of application:

The Mamdani method is currently applied in process control, robotics, and other expert systems. It is particularly suitable for the execution of command-and-control actions of an operator. It leads to good results that are often close to those of a human operator, eliminating the risk of human error. Thus, it has been and is successfully used in the control of the activity on industrial platforms.

The Mamdani-type control is simpler than the others considered as standard and requires a much shorter development cycle when the rules can be expressed simply, for the simple reason that it is not necessary to develop, analyze, and implement a mathematical model. The method is preferred if the process to be controlled is governed by nonlinear laws or includes imprecise parameters or strong perturbations (in the case of an imprecise model). In the classical approaches, when the mathematical model contains some nonlinear terms, are linearized and simplified, assuming the occurrence of errors with low values, while a fuzzy approach often allows control of larger areas of error. The nonlinearity of the Mamdani fuzzy control can have a favorable influence in the case of transient phenomena. Thus, it can replace the classic control in order to ensure a quick response. In conventional controllers, with the increase in the response speed, the overgrowth parameter also increases. In general, fuzzy controllers with pronounced nonlinear characteristics generate smaller overgrowths compared to conventional PID control, but oscillations can still be observed after the set time. This behavior can be very difficult to remove. However, by choosing the correct fuzzy sets and rule sets, fuzzy controllers can have a perfectly linear feature, thus completely replacing standard controllers, which can only reserve the role of providing correction data.

This limitation of the regulating hypersurface can be reduced by defining several fuzzy sets on the input universe. The disadvantage of this method, however, is the increase in the number of rules and the complexity of the system (the rule base becomes more difficult to meaning). Another possibility of reduction is the use of fuzzy rules of multiple weighted consequences.

Presentation of the practical application:

The configuration of this proposed controller is as follows:

- Inputs: temperature e, and humidity de
- Output: command c

Fuzzification will be performed according to the following formula:

- for input e, seven MF: NG, NM, NS, Z, PS, PM, and PG
- for the input de, five MF: NG, NS, Z, PS, and PG
- for output c, five MF: NG, NS, Z, PS, and PG

The language tags used are traditional:

- NG — negative great;
- NM — negative medium;
- NS — small negative;
- Z — zero/zero;
- PS — positive small;
- PM — positive medium;
- PG — positive great.

Only normalized variables will be used, and adaptation to the driven process is achieved by three scaling factors: the scaling factor for input FSe, the scaling factor for the FSde input derivative, and the scaling factor for the command FSc.

The proposed basis of rules is as follows:

1. If (e is NG), then (c is PG)
2. If (e is NM) and (de is NG), then (c is PG)
3. If (e is NM) and (de is NS), then (c is PG)
4. If (e is NM) and (de is Z), then (c is PS)
5. If (e is NM) and (de is PS), then (c is PS)
6. If (e is NM) and (de is PG), then (c is Z)
7. If (e is NS) and (de is NG), then (c is PG)
8. If (e is NS) and (de is NS), then (c is PS)
9. If (e is NS) and (de is Z), then (c is PS)
10. If (e is NS) and (de is PS), then (c is Z)

11. If (*e* is NS) and (*de* is PG), then (*c* is NS)
12. If (*e* is Z) and (*de* is NG), then (*c* is PS)
13. If (*e* is Z) and (*de* is NS), then (*c* is PS)
14. If (*e* is Z) and (*de* is Z), then (c is Z)
15. If (*e* is Z) and (*de* is PS), then (*c* is NS)
16. If (*e* is Z) and (*de* is PG), then (*c* is NS)
17. If (*e* is PS) and (*de* is NG), then (*c* is PS)
18. If (*e* is PS) and (*de* is NS), then (*c* is Z)
19. If (*e* is PS) and (*de* is Z), then (*c* is NS)
20. If (*e* is PS) and (*de* is PS), then (*c* is NS)
21. If (*e* is PS) and (*de* is PG), then (c is NG)
22. If (*e* is PM) and (*de* is NG), then (*c* is Z)
23. If (*e* is PM) and (*de* is NS), then (*c* is NS)
24. If (*e* is PM) and (*de* is Z), then (*c* is NS)
25. If (*e* is PM) and (*de* is PS), then (*c* is NG)
26. If (*e* is PM) and (*de* is PG), then (*c* is NG)
27. If (*e* is PG), then (*c* is NG)

Figure 3.12 shows the graphical interface for the fuzzy control architecture for greenhouse applications, and Figure 3.13 shows the ways to implement the membership function for the two temperature and humidity inputs.

Fuzzy-mean defuzzification is a weighted sum and can be considered a particular case of the CoA method. When the center of area (CoA) method is applied, the result of the defuzzification is a nonlinear function that has the inputs of the regulator as variables (in most cases). The fuzzy controller is itself a nonlinear controller. The aggregation with the max operator in combination with the center of maximum (CoM) method. Thus, the CoM defuzzification method in combination with aggregation operators other than the sum operator will enter nonlinearities in the output function of the controller. Figure 3.14 shows the nonlinear controller for the CoA method using singleton membership functions.

For the implications that are subject to the classical conjunction, the aggregation operator is the disjunction. For the implications that are subject to the implication, the classic aggregation operator is the conjunction. Figure 3.15 shows the aggregation of two fuzzy rules using the operator max. The min operator is used as an implication function for the two rules in Figure 3.15. The inference is a procedure that allows the determination of new knowledge starting from the concrete data of the problem to be solved.

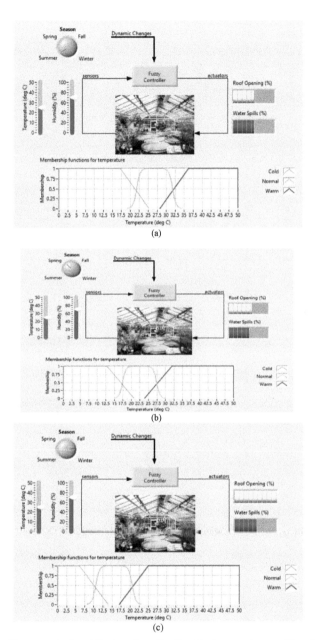

Figure 3.12 Fuzzy control architecture for adjusting the temperature and humidity from the greenhouse: (a) summer; (b) spring; (c) fall.

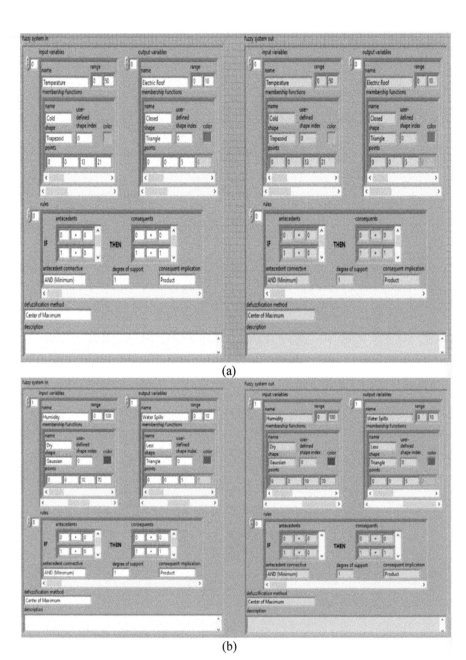

Figure 3.13   Membership function for two inputs: (a) temperature and (b) humidity.

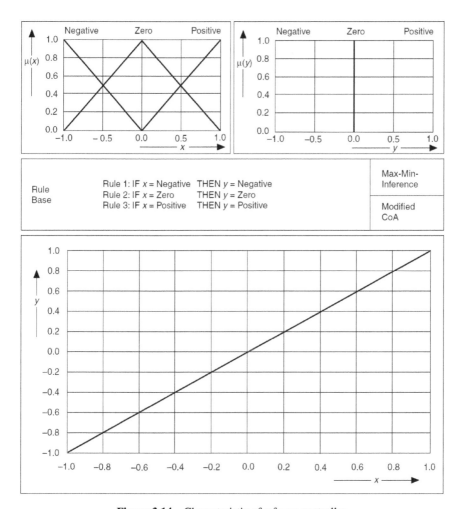

Figure 3.14 Characteristic of a fuzzy controller.

The implementation of the fuzzy controller based on the theoretical approach requires the discretization of fuzzy relationships for their storage in computer memory. To minimize errors due to discretization, the discretization step must be chosen as small enough. If the error is small and the error change frequency is high, then it reduces the control that can be part of a fuzzy PI-type controller. Considering the specialized literature, it can be stated that the most used are the implications of T. This behavior results from output characteristics, as shown in Figure 3.15.

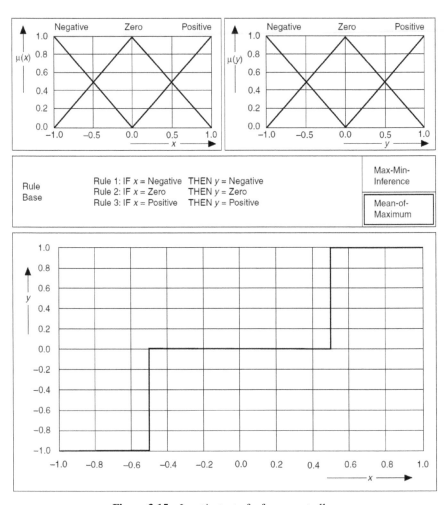

Figure 3.15 Input/output of a fuzzy controller.

Aggregation is the operation used to combine the fuzzy sets obtained as the output of each fuzzy rule from the rule base, known as partial conclusions, to obtain the fuzzy set that represents the response of the system with fuzzy logic to a given value of the transient input: this can be max, sum (algebraic sum), or prob-or, and, by default, it is max. Figure 3.16 shows how the controller characteristic can be changed if we change the rule base from the previous example by introducing the following rules:

- Rule 1: IF x = negative, THEN y = negative
- Rule 2: IF x = zero, THEN y = positive
- Rule 3: IF x = positive, THEN y = negative

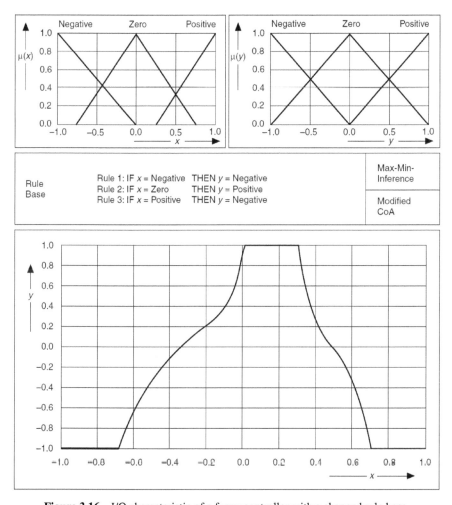

Figure 3.16 I/O characteristic of a fuzzy controller with a changed rule base.

Using a minimum number of memberships of fuzzy controller output will lead to restrictions on the control hypersurface, especially if the inputs are normal and form a fuzzy partition and if the inputs are numeric (Figure 3.17).

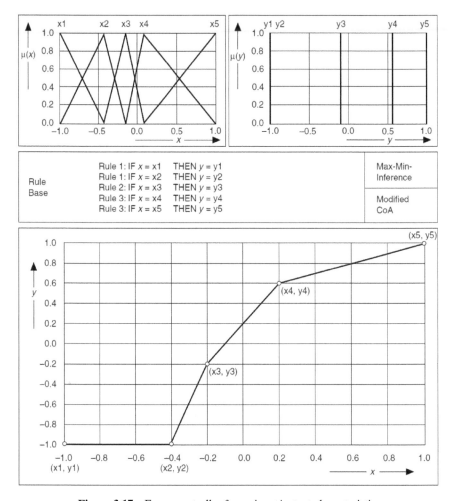

Figure 3.17   Fuzzy controller for an input/output characteristic.

This limitation is caused by using a single language tag defined for the controller output when applying the fuzzy rule. Figure 3.18 shows the input/output characteristic for a fuzzy controller using two inputs.

The AND method allows the selection of the mathematical expression used for the AND operator from the premise of the fuzzy rule: this can be min or prod (product), and, by default, it is min, as shown in Figure 3.19.

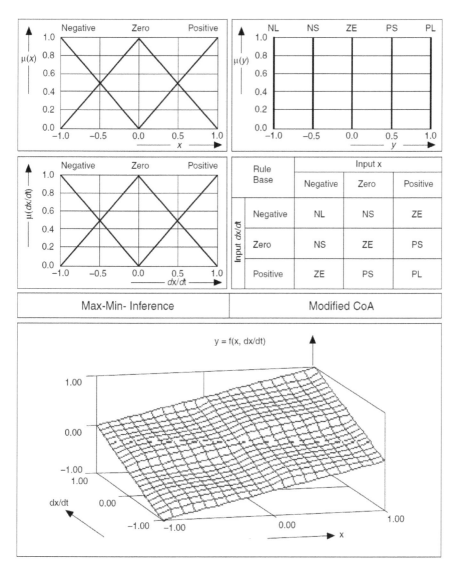

Figure 3.18 Input/output characteristic field of a fuzzy controller using two inputs.

3.2.6 Discussion

After the implementation of the two regulators, it can be easily seen that the simulations do not raise any computational problems at all; they take place practically instantly, without the risk of blockages. FACCT files can

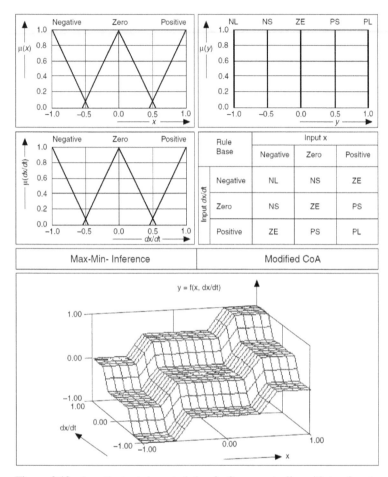

Figure 3.19   Input/output characteristic of a fuzzy controller with two inputs.

be compiled by fuzzy cells. The cells are implemented on the structure of an INTEL Quark microcontroller C1000. Unlike the use of the *.fis file in which blockages occur for certain values of the control system parameters, we can now try any combination of control system parameters, for example, increasing the output scaling factor from 10 to 25. There is a great similarity between the two answers over time, an argument in favor of using the simpler variant, with 25 rules. But the number of rules itself is more or less relevant than how these rules are formulated and adjusted. As can be seen, the answers support many improvements. The major difference between fuzzy regulators and conventional regulators is the much larger number of adjustment mechanisms that can influence performance. This gives them a fundamental advantage, especially in difficult applications or about which we do not have enough knowledge to treat them optimally, but, on the other hand, it requires a more laborious adjustment. If in the case of a linear PD controller there are only two adjustment parameters, to which we can add the scaling factors at the input, in the case of the fuzzy-interpolating controller, we can identify at least the following adjustment parameters: the scaling factor at the output FSc, input fuzzification comprising 25 interpolation nodes, and output fuzzification with five values. In the case of an ordinary fuzzy system, the forms of the membership functions also intervene, through the possibility of using the Gaussian, sigmoidal functions. Example of distributed fuzzy control using fuzzy cells operating based on an application written in FACCT language in distance control applications between two moving objects and in measuring the distance to a target.

3.3 Conclusion

Intelligent distributed control can be defined and used: fuzzy sensors – as a representation of input elements, and fuzzy actuators – which act in the real world, depending on the input data they receive, to which we add the input elements fuzzy inference, which achieves distributed fuzzy logic. Thus, the system makes new fuzzy subsets from the fuzzy subsets received from the input elements. Fuzzy components can be integrated for different applications into a compact model of components, called the fuzzy cell. There is also the possibility to create virtual components. The main directions for further research are the following: creating a compiler for the FACCT language that allows implementation with the help of digital or analog fuzzy processors of different distributed fuzzy control configurations. In this way, the language described in this chapter will be able to be used in the development of fuzzy components with intelligence systems on the structure of configurations with microprocessors or analog circuits.

Acknowledgement

Many thanks for the support from the Faculty of Transportation management, and the Department of Telematics and Electronics for Transports at the University Politehnica of Bucharest, Romania, and for allowing the use of research from Artificial Intelligence and Telematics for Transports Laboratories.

References

[1] D. Akif, I. Uzeyir and C. Mehmet, 'Design of a Distributed Control System with Fuzzy Logic Controller and PLC in Wireless Sensors Network based Industrial Environments and Monitoring the System with RFID.' International Journal of Engineering Research and Development, vol. 11, pp. 173-191, January 2019, 10.29137/umagd.396400.

[2] E. Helbert, M. Ivan, L. Hilario and D. Guzman, 'Proposal of an Adaptive Neurofuzzy System to Control Flow Power in Distributed Generation Systems.' Vol. 2019, article ID. 1610898, pp. 16, Wiley – Hindawi, DOI: 10.1155/2019/1610898.

[3] D. Dufton and C. Collier, Fuzzy Logic filtering of radar reflectivity to remove non-meteorological echoes using dual polarization radar moments, Atmospheric Measurement Technique, 2015, Atmos. Meas. Tech., 8, 3985–4000, 2015, www.atmos-meas-tech.net/8/3985/2015/, doi:10.5194/amt-8-3985-2015.

[4] B. Abro, K. Hyun, C. Myeong, O. Rymduck, A. Akmal and S. Heung, 'Application of fuzzy logic for problem of evaluating states of a computing system, article.' Journal Applied Science MDPI, published 26 July 2019, Appl. Sci. 2019, 9, 3021; doi:10.3390/app9153021.

[5] K. Mason, M. Duggan, E. Barrett and J. Duggan, 'Howley, E. Predicting host CPU utilization in the cloud using evolutionary neural networks.' Future Gener. Comput. Syst. 2018, vol. 86, pp. 162–173.

[6] J. Antonio, C. Juan, T. Alicia, 'A distributed clustering algorithms guided by the base station to extend the lifetime of wireless sensors networks.' Journal Sensors, 2020, vol. 20, pp. 2312; doi:10.3390/s20082312.

[7] Q. Jiang, X. Jin, J. Hou, S. Lee and S. Yao, 'Multi-Sensor Image Fusion Based on Interval Type-2 Fuzzy Sets and Regional Features in Nonsubsampled Shearlet Transform Domain.' IEEE Sens. J. 2018, vol. 18, pp. 2494–2505.

[8] E. H. Mamdani, 'Application of fuzzy algorithms for control of simple dynamic plant.' Proc. Inst. Electr. Eng. 1974, vol. 121, pp. 1585–1588.

[9] J. J. Jassbi, P. J. A. Serra, R. A. Ribeiro and A. A. Donati, 'Comparison of Mandani and Sugeno Inference Systems for a Space Fault Detection Application.' In Proceedings of the 2006 World Automation Congress, Budapest, Hungary, 24–26 July 2006; pp. 1–8.

[10] Y. Zhang, J. Wang, D. Han, H. Wu and R. Zhou, 'Fuzzy-logic based distributed energy-efficient clustering algorithm for wireless sensor networks.' Sensors 2017, vol. 17, pp. 1554.

[11] A. J. Yuste-Delgado, J. C. Cuevas-Martinez and A. Triviño-Cabrera, 'EUDFC-Enhanced Unequal Distributed Type-2 Fuzzy Clustering Algorithm.' IEEE Sens. J. 2019, vol. 19, pp. 4705–4716.

[12] K. Thangaramya, K. Kulothungan, R. Logambigai and M. Selvi, 'Ganapathy, S.; Kannan, A. Energy aware cluster and neuro-fuzzy based routing algorithm for wireless sensor networks in IoT.' Computer Net. 2019, vol. 151, pp. 211–223.

[13] J. X. Xu, Z. Q. Guo and T. H. Lee, 'Design and implementation of a Takagi-Sugeno-Type fuzzy logic controller on a two-wheeled mobile robot.' IEEE Trans. Ind. Electron. 2013, vol. 60, pp. 5717–5728.

[14] C.-H. Huang, W.-J. Wang and C.-H. Chiu, 'Design and implementation of fuzzy control on a two-wheel inverted pendulum.' IEEE Trans. Ind. Electron. 2011, vol. 58, pp. 2988–3001.

[15] H. Chih and F. Ya, 'Design of Takagi-Sugeno fuzzy control scheme for real word system control.' Journal Sustainability 2019, vol. 11, pp. 3855; doi:10.3390/su11143855

[16] X. Su, Y. Wu, J. Song and P. Yuan, 'A Fuzzy Path Selection Strategy for Aircraft Landing on a Carrier.' Appl. Sci. 2018, vol. 8, pp. 779.

[17] P. G. Zavlangas, S. G. Tzafestas and K. Althoefer, 'Fuzzy Obstacle Avoidance and Navigation for Omnidirectional Mobile Robots; European Symposium on Intelligent Techniques: Aachen, Germany, 2000.

[18] M. Nadour, M. Boumehraz, L. Cherroun and V. Puig Cayuela, ' Hybrid type- fuzzy logic obstacle avoidance system based on horn-schunck method.' Electroteh. Electron. Autom. 2019, vol. 67, pp. 45–51.

[19] H. Jahanshahi, M. Jafarzadeh, N. N. Sari, V. T. Pham, V. V. Huynh and X. Q. Nguyen, 'Robot motion planning in an unknown environment with danger space.' Electronics 2019, vol. 8, pp. 201.

[20] J. Lin, J. Zhou, M. Lu, H. Wang and A. Yi, 'Design of Robust Adaptive Fuzzy Controller for a Class of Single-Input Single Output (SISO) Uncertain Nonlinear System.' Math. Probl. Eng. 2020, 2020, 6178678.

[21] F. Rossomando, E. Serrano, C. Soria and G. Scaglia, 'Neural Dynamics Variations Observer Designed for Robot Manipulator Control Using a Novel Saturated Control Technique.' Math. Probl. Eng. 2020, 2020, 3240210.

[22] A. Chatterjee and K. Watanabe, 'An adaptive fuzzy strategy for motion control of robot manipulators.' Soft Comput. 2005, vol. 9, pp. 185–193.

[23] R. N. Bandara, Gaspe, S., Fuzzy logic controller design for an Unmanned Aerial Vehicle.' In Proceedings of the IEEE International Conference on Information and Automation for Sustainability, Galle, Sri Lanka, 16–19 December 2016; pp. 1–5.

[24] M. Prakash and K. Jajulwar, 'Design of adaptive fuzzy tracking controller for Autonomous navigation system.' Int. J. Recent Trend Eng. Res. 2016, vol. 2, pp. 268–275.

[25] H. Xue, Z. Zhang, M. Wu and P. Chen, 'Fuzzy Controller for Autonomous Vehicle Based on Rough Sets.' IEEE Access 2019, vol. 7, pp. 147350–147361.

[26] G. C. Karras, 'Fourlas, G. K. Model Predictive Fault Tolerant Control for Omni-directional Mobile Robots.' J. Intell. Robot. Syst. 2020, vol. 97, pp. 635–655.

[27] R. Siegwart, I. R. Nourbakhsh and D. Scaramuzza, 'Introduction to Autonomous Mobile Robots.' MIT Press: Cambridge, MA, USA, 2011.

[28] T. Mac, C. Copot, R. De Keyser, T. Tran and T. Vu, 'MIMO fuzzy control for autonomous mobile robot.' J. Autom. Control. Eng. 2016, vol. 4, pp. 65–70.

[29] M. V. Bobyr, S. A. Kulabukhov and N. A. Milostnaya, 'Fuzzy control system of robot angular attitude.' In Proceedings of the 2nd International Conference on Industrial Engineering, Applications and Manufacturing, Chelyabinsk, Russia, 19–20 May 2016; pp.1–6.

[30] K. Li, X. Zhao, S. Sun and M. Tan, 'Robust target tracking and following for a mobile robot.' Int. J. Robot. Autom. 2018, vol. 33, no. 4.

[31] A. Pandey, S. Pandey and D. Parhi, ' Mobile robot navigation and obstacle avoidance techniques: A review.' Int. Robot. Autom. J. 2017.

4

Evaluating the Effectiveness of Enterprises' Digital Transformation by Fuzzy Logic

Oleg Kaminsky[1], Viktor Koval[2], Julia Yereshko[3], Nataliia Vdovenko[4], Mykhailo Bocharov[5], and Yigit Kazancoglu[6]

[1]Department of Computer Mathematics and Information Security, Kyiv National Economic University named after Vadym Hetman, Ukraine
[2]National Academy of Sciences of Ukraine, Ukraine
[3]Department of Economic Cybernetics, National Technical University of Ukraine "Igor Sikorsky Kyiv Polytechnic Institute," Ukraine
[4]National University of Life and Environmental Sciences of Ukraine, Ukraine
[5]Research Laboratory Moral and Psychological Support of Troops (Forces) Activity Department, National Defense University of Ukraine named after Ivan Cherniakhovskyi, Ukraine
[6]Department of Logistics Management, Yasar University, Turkey

Abstract

The study analyzes the trends in the digital economy, as well as some aspects of an enterprise's digital transformation in modern conditions, and also determines the criteria and directions for such a transformation according to the developed ecosystem model for the enterprise's digital transformation based on fuzzy logic. The proposed model takes into account the subjectivity resulting from the measurement and setting of transformation goals by using various standardized rules. The chosen research methodology made it possible to identify existing and new patterns of interaction between agents in the context of digital transformation of the organizational and economic system of an enterprise and to determine that an enterprise digital transformation strategy should have a broader focus and include information architecture reengineering, the creation of digital platforms, and introduction of additive technologies. Choosing a digital transformation strategy based on summary performance indicators and fuzzy logic methods will allow companies to achieve long-term strategic goals and determine their place in the digital market.

Keywords: Digital economy, digital transformation, fuzzy logic, performance evaluation, key performance indicators, ecosystem, innovations.

4.1 Introduction

The modern world is characterized by the dynamic pace of the formation of new information and computer technologies and means of network interaction, which have led to the emergence of new scientific terms "digital goods" and "digital economy." The digital economy is characterized by the fact that most of the gross domestic product is provided by the production, processing, storage and dissemination of digital goods, and knowledge, and is carried out through platforms such as the Internet, as well as through mobile and social networks. The digital economy reflects the process of transition from the third to the fourth industrial revolution and leads to the transformation of relations between economic entities in many industries.

The development of the digital economy fundamentally changes the interaction of economic entities, production processes, and cross-border trade. The structural crisis of the economy caused by the COVID-19 pandemic could potentially be overcome by introducing new technologies that create new product opportunities and assimilation, which provides a breakthrough in improving its efficiency and future transition to a new stage of growth. However, doing business within the digital economic paradigm determines the need for significant changes in the organizational practice, structure, and business processes of enterprises and economic entities. For businesses in the digital economy, it is necessary to create an effective planning and control system to support their digital transformation strategies.

Huawei's Global Industry Vision analytical report [1] shows that by 2025, 97% of the world's largest enterprises will use advanced digital technologies in their production and business processes; the level of smart robots application in the field of housing and communal services will reach 14%, and the percentage of industrial enterprises using AR/VR technologies will increase up to 10%; the number of intelligent personal digital assistants, in turn, will reach 90%.

Industrial robots would work alongside people in production: as the analysis shows, for every 10,000 employees, it is planned to deploy about 100 robots; enterprises will also efficiently use up to 86% of their data; up to 85% of business software applications for enterprises will become cloud services, and 5G networks will cover 58% of the world's population; the total amount of global data, including enterprise data, will reach 180 Zbytes [1]. The Fourth Industrial Revolution paradigm and its consequences for the global economy were explored in [2] by Professor Klaus Schwab, the founder and chairman of the World Economic Forum (WEF), who described the phenomenon as a "fusion of technologies that blurs the physical, digital and biological spheres." That is, the basis for future transformation will be the development of cyber−physical industrial systems.

There are processes in the economy that can be in a state of uncertainty. Modeling such processes with precise mathematical methods could be very difficult or even impossible. To find a way out of such situations, Professor Lutfi Zadeh (1965) developed a method of fuzzy logic of a continuum with many values [3]. Thus, due to the high level of the subjectivity of management models in the context of digital transformation, it is necessary to apply a fuzzy methodology going forward.

The basic concept of mathematics is a set or collection of objects. Gradually, however, it became clear that most human knowledge and connections to the outside world include constructions that cannot be considered sets in the classical sense. Rather, they should be considered "fuzzy sets," i.e., classes with blurred boundaries, when the transition from belonging to a certain class to non-belonging occurs sharply but is gradual.

Based on this, L. Zadeh formulated the thesis that the logic of human reasoning is based not on classical ambiguous or even polysemantic logic but on logic with fuzzy truth values, fuzzy connections, and fuzzy inference rules [4]. And, of course, it was necessary to understand how it is possible to operate with fuzzy sets, given the limitations of the methods of classical mathematics.

Finally, and most importantly, the challenge is to develop new methods to systematically deal with ambiguities. Thanks to the research of L. Zadeh, new research opportunities have opened up in economics, sociology, political science, linguistics, and the fields of operations research, management theory, etc. [5].

Currently, fuzzy logic has become widespread as a tool of analysis. It is used to develop various data mining methods that are widely used to study economic processes. As an example, we can point out the verified methods of fuzzy logic in economics, in such a wide range as:

- construction of neural networks based on soft calculations, as well as their adaptive variants;
- database processing using fuzzy queries;
- fuzzy associative rules;
- fuzzy cognitive maps, which can be used to model the conditions for the functioning of an enterprise.

For example, in a study by B. Diaz and A. Morillas [6], the use of fuzzy methods for modeling and forecasting paid employment in Spain using the Young algorithm is proposed. The paper also explores the problems that can arise in the analysis of industrial sectors and clusters due to a typical outlier when using multivariate data.

The study by J. Bih [7] defines the concept of a linguistic variable as an important aspect of fuzzy logic, the values of which are words or sentences in natural language; it is emphasized that the use of fuzzy logic algorithms in a digital and integrated economy is undeniably relevant. They automate decision-making processes, perform multidimensional analysis of large amounts of data, and learn from mistakes. Fuzzy logic, with its huge potential, allows for high decision-making productivity by integrating it with big data management, artificial intelligence, neural networks, and more.

Researchers Nasiri and Darestani [8] determined in their work that the business sector needs optimal, fast, and efficient quality control tools for its development. Based on this premise and taking into account that there is not always accurate data or under conditions of uncertainty, the use of fuzzy logic algorithms allows process modeling.

4.2 Results

As a new stage in the development of society, the information society began to take shape as a result of the revolutionary impact of information technologies, and its modern structure was formed under the influence of globalization processes.

We can single out the period of formation and development of the information (digital) economy as the main type of economic system of the information society. This period can be defined from 1996 (the year of the first commercial projects on the Internet) to the present. However, those dates are indicative and may be the subject of discussion. At the heart of the new (digital) economy is the process of turning digital goods and services into mass production.

It becomes necessary to develop digital entrepreneurship to solve the problems associated with the structural transformation of the economy, and to support the period of such a transformation of the level of business activity, the emergence of new jobs, the development of financial processes, etc. As shown in the research [10], the transition of entrepreneurial activity into the digital plane is becoming a factor contributing to the maintenance of sustainable development trends in the transition period.

In this case, we consider a digital product as an entity (good or service) that has value (for example, software or a data package for 3D printing), the cost of making each copy of which is zero.

Thus, digital transformation is a transition to the production of digital products and business models (management, logistics, labor organization, etc.) using digital platforms. The dynamics of the development of the digital economy are at a high pace, due to the flexibility and scalability of the infrastructure of the digital market, the specifics of the life cycle of digital goods, and innovative activity.

According to [9] and [11], the paradigm of a digital economy includes:

- ancillary infrastructure (hardware and software, infrastructure, telecommunications, computer networks, etc.);
- E-business (conducting economic activities and other business processes through computer networks);
- E-commerce (selling goods over the Internet).

Information technology and the Internet of Things are reinforcing the trend of innovative economic development, offering new uses for products and new segments of consumers of business models such as the subscription model and the "as a service" model. The industrial Internet of Things allows companies to plan their activities more accurately and increase revenue by providing a service or product over a period of time (subscription models).

The main directions of the modern economy megatrends development show that the transition to a digital economy, the development of the knowledge economy, intellectual capital, and additive manufacturing create conditions for the growth of digital entrepreneurship within the framework of the "Industry 4.0" concept, the basic components of which are given in Table 4.1.

Table 4.1   Industry 4.0 paradigm and its components.Source: Developed by the authors.

Categories	Technologies
Integration of vertical and horizontal value chains of the digital economy	Human—machine and M2M interfaces Cloud computing Mobile and portable devices Internet of Things (IoT)
Digitization of industrial products, equipment, and services	Virtual reality/portable devices (implants) Interaction with clients through social networks and profiling of clients by machine learning methods Multidimensional analysis of big data and genetic algorithms Smart sensors Geolocation technologies
New digital business models and business processes	Additive economy and 3D printers Digital platforms Global digital commerce Digital finances and banking Digital governance

Doing business in the digital economy requires restructuring the organizational structure of the enterprise and introducing new business models. It is necessary to switch to the use of cloud computing, develop methods for managing company data, introduce new forms of legislation, build our digital platforms, and retrain specialists, due to the demand for digital skills. Marketer, copywriter, sales manager, or HR — every employee must be able to manage digital data.

The main directions and trends of digital transformation of enterprises are shown in Figure 4.1.

Providing an assessment of the effectiveness of a company's digitalization in general and the digital ecosystem it has created in particular, it is necessary to identify the basis for analysis.

The essence of the "ecosystem" is related to the relationship between its subjects and the environment in which they exist, which is also characterized by the heterogeneity of its components. Subsequently, the term "business ecosystem" was proposed and the concept of ecosystems has developed significantly in socio-economic research.

The problem of timeliness and compliance of innovations with the needs of economic development encourages rethinking the transformation process on the path to innovation.

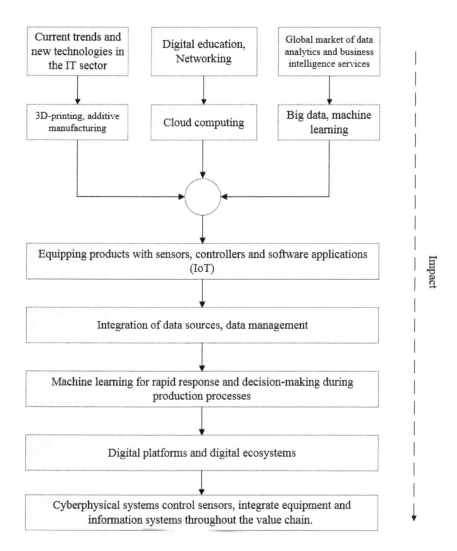

Figure 4.1 Trends in the digital transformation of enterprises
Source: Developed by the authors.

Based on research on the patterns of development, as well as the reasons for the incremental intellectualization of entrepreneurial activity, we propose an ecosystem model of digital transformation of the enterprise based on the determination of its specific need for innovation (Figure 4.2).

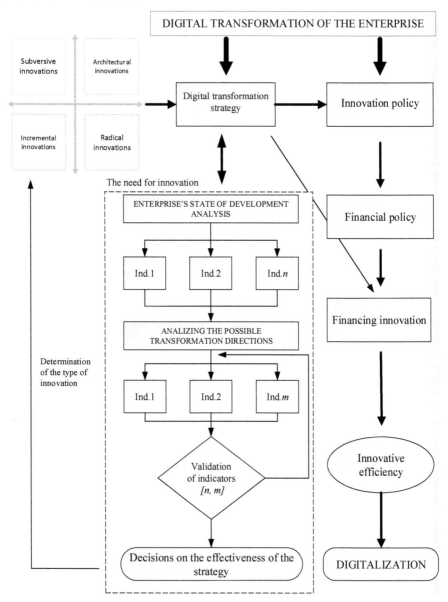

Figure 4.2 Ecosystem of digital transformation of the enterprise.
Source: Developed by the authors.

Modern scientific research uses different definitions of ecosystems; in the framework of this work, by ecosystem, we mean the unity of products/services of the company, developed and implemented by it for different areas and markets. But the ecosystem includes not only the company's products or services; this concept is more global. The ecosystem also includes the company's business processes, its counterparties, and end-users of goods/services.

This set operates exclusively in digital format, which requires a full-fledged digital platform, which is why the pioneers in creating digital ecosystems are IT companies.

Signs of the enterprise's digital ecosystem can be identified as follows:

- unified interface for access to products and services and ways to build them regardless of the product environment faced by the customer;
- unified means to identify users;
- seamless data transfer;
- seamless loop of data movement between frontal environments.

Creating a digital ecosystem has a number of goals that can be attributed to it:

- increase in market share of the enterprise;
- attracting more customers;
- forming a constant loyal audience of consumers (brand);
- enterprise activities diversification;
- maintaining market positions, etc.

Research shows that the level of profit, sector, brand, and leadership in the field of innovation does not automatically provide the company with sustainable development of the digital ecosystem. Untimely development and/or imitation of innovation can have a significant impact on economic security. Therefore, it is important to introduce methods for assessing the digital transformation of the enterprise, which will ensure the implementation of innovative development strategies while maintaining the appropriate level of financial and economic stability.

To assess the effectiveness of the enterprise digital ecosystem, we need to determine both the directions of transformation and evaluation indicators. We propose to use the indicators of digital transformation efficiency, which are given in Table 4.2.

Various tools, including fuzzy logic ones, can be used to assess the feasibility and effectiveness of digitization in achieving a company's strategic

Table 4.2   Key efficiency indicators of the enterprise digital ecosystem.

Enterprise development strategy	Performance indicators
Production strategy	Sales dynamics indicator
	Production volume dynamics indicator
	Labor productivity dynamics indicator
	The level of additive technologies implementation
	Service quality indicator
Financial strategy	Asset growth indicator
	Capital growth indicator
	Loan increase indicator (share of loan capital)
	Asset turnover indicator
	Profitability growth indicator
	Financial stability indicator
	Indicator of the investment dynamics (in digital technologies)
Organizational strategy	Indicator of the staff turnover dynamics
	Percentage of staff involvement
	Innovation index
	Job satisfaction index
	Index of staff creativity
Transformation strategy	IT infrastructure spending indicator
	Digital platform cost indicator
Branding strategy	Enterprise reputation indicator
	Market value dynamics indicator
	Innovative development dynamics indicator

Source: Developed by the authors.

goals. Businesses can evaluate digital transformation based on ratios "plan – fact," calculated at several hierarchical levels of analysis.

The classification of the index digital transformation strategic efficiency can be defined as a fuzzy subset.

One such approach is L. Zadeh's fuzzy logic. In his work [3], the concept of a set is extended by the assumption that the membership function of an element to a set can take any values in the interval [0, 1], and not just 0 or 1.

From the indicators of the economic condition of the enterprise, we chose the indicators that are given in Table 4.2. When choosing indicators, the criterion is the contribution to the company's digital transformation process. The formulation of this criterion is necessary to calculate the performance indicators of the transformation strategy.

The lower level of the model for evaluating the effectiveness of a company's digital transformation (Figure 4.3) consists of the various economic indicators' values.

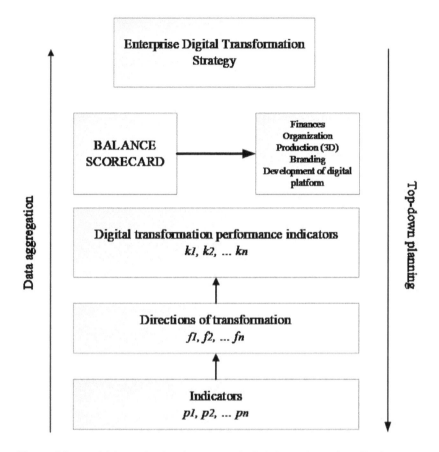

Figure 4.3 Model for evaluating the company's digital transformation effectiveness.
Source: Developed by the authors.

The values of the indicators are determined by the company using various monitoring systems and measurement points. To be able to effectively assess the directions of transformation, it is necessary to aggregate indicators on certain aspects. As a method of aggregation for enterprises, grouping by areas of digital transformation strategy is recommended. At the second level of the model, groups are formed according to the directions of the digital transformation strategy. At the third level of the model, the values of indicators are aggregated.

At the next level, the grouped KPI values are aggregated according to the balanced scorecard (BSC) methodology. The upper level is the strategic

Table 4.3   Digital transformation effectiveness evaluation.

Grades	Indicators values
Very inefficient	If $W_j < 0.95$
Unacceptable	If $W_j \in [0.95; 1)$
Satisfactory	If $W_j \in (1; 1.05)$
Acceptable	If $W_j \in (1.05; 1.1]$
Very effective	If $W_j > 1.1$

level, where the strategic index of transformation efficiency is calculated. Using this index, you can assess the strategic efficiency of the enterprise in a digital economy. The application of the balanced scorecard (BSC) system makes it possible to assess the economic consequences of digital transformation and, therefore, the effectiveness of planning and implementation of the transformation program from different strategic points of view.

At the level of groups of transformation directions and BSC (system of balanced indicators), different aggregates have different weights, as shown in Figure 4.3.

Thus, a BSC-based model for evaluating the effectiveness of digitalization of the studied company is built. A fuzzy methodology is used to include the high subjectivity of the control model.

Classification of indicators according to different norms of standardization indicates that the indicator can be assigned to another group with the same coefficient value (subjective classification based on plan-fact differences in the planning period determined by the enterprise, and based on five evaluation classes) and threshold values (classified indicators that can be assigned to another class when the norm changes) (Table 4.3).

The function used for classification is given in eqn (4.1):

$$F_i = \frac{\sum \frac{Z_{ij}}{P_i} W_j}{T}, \qquad (4.1)$$

where Z is the actual indicator value, P is the predefined target value, i and j are numbers of researched elements, T is the number of tested items belonging to the BSC directions, and W_j is the indicator weight value.

Businesses appoint five different classes to assess the effectiveness of their indications (Table 4.3).

The fuzzy function used for classification is presented in Figure 4.4.

The company will be able to analyze the effectiveness of digital transformation in five classification categories. The current feature allows using the controlling system for feedback.

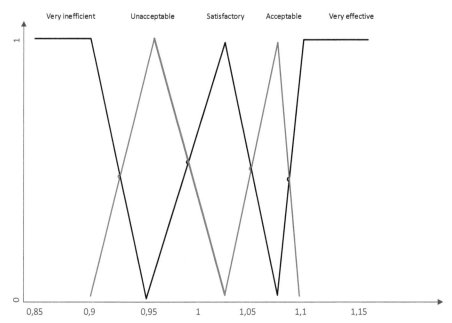

Figure 4.4 The fuzzy function of evaluating the company's digital transformation effectiveness.

Source: Developed by the authors.

For businesses in a digital economy, it is necessary to create an effective planning and control system to support digitization strategies. Controlling will be a considerable source of information for decision support processes for digital transformation at all levels. This will help to adapt to the dynamic changing environment of the digital society.

4.3 Conclusion

The study presents a model based on fuzzy logic, which allows companies to effectively control the effectiveness of digital transformation.

This model, in contrast to standard models for calculating efficiency, is used under conditions of uncertainty and takes into account the possible dimensions of the objectives of the strategy. The level of subjectivity of the goals of the strategy and its indicators are considered by the model by the evaluation method using various standardized rules. This allows you to build a digital transformation strategy that takes into account various contexts regarding goals, the influence of external factors, and the environment. Using such a model, one can evaluate unique transformation strategies according to different business areas.

In a deconstructive economic environment, the assessment of enterprises always depends on the context of the study. The war in Europe and the COVID-19 pandemic have affected the economy, as well as significantly affected the real assessment of companies and their place in the digital market. Thus, by evaluating the business by following the internal goals of the enterprise, you can achieve more relevant information content and clearer support for management decisions.

In times of crisis, such as war, people are acutely aware of uncertainty due to a number of factors: employment problems, lack of or limited funds, limited commodities, the impossibility of long-term planning, and constant fear for their lives and the lives of loved ones.

Thus, the outcome of the model evaluates the actual effectiveness of the transformation strategy in various contexts compared to the expectations from its implementation.

The model focuses on both short-term and long-term project analyses as it standardizes transformation performance through plan-to-fact analysis. Thus, it can display the strategic performance of a digital transformation project in one metric.

References

[1] Global Industry Vision 2025: https://www.huawei.com/minisite/giv/en/index.html

[2] K. Schwab, 'The Fourth Industrial Revolution. What it Means and How to Respond'. 2015. URL: https://www.weforum.org/agenda/2016/01/the-fourthindustrial-revolution-what-it-means-and-how-to-respond/.

[3] L. A. Zadeh, 'Fuzzy sets'. Inf. Control 1965, 8, 338–353.

[4] L. A. Zadeh, 'Is there a need for fuzzy logic?' Inf. Sci. 2008, 178, 2751–2779.

[5] F. Zadeh, 'My Life and Travels with the Father of Fuzzy Logic'. Albuquerque, New Mexico, USA: TSI Press, 1998.

[6] B. Díaz, A. Morillas, 'Some Experiences Applying Fuzzy Logic to Economics'. In: Seising R., Sanz González V. (eds) Soft Computing in Humanities and Social Sciences. Studies in Fuzziness and Soft Computing, 2012 vol 273. Springer, Berlin, Heidelberg. https://doi.org/10.1007/978-3-642-24672-2_19

[7] J. Bih, 'Paradigm Shift—An Introduction to Fuzzy Logic'. IEEE Potentials, 2006, 25, 6-21. https://doi.org/10.1109/MP.2006.163502 1

[8] M. Nasiri, S. A. Darestani, A Literature Review Investigation on Quality Control Charts Based on Fuzzy Logic. International Journal of Productivity and Quality Management, 2016, 18, 474-498. https://doi.org/10.1504/IJPQM.2016.077778

[9] W. H. Davidow, M. S. Malone, 'The Virtual Corporation: Structuring and Revitalizing the Corporation for the 21st Century'. New York: HarperCollins Publishers, 1993.

[10] O. Y. Kaminsky, Y. O. Yereshko, S. O., Kyrychenko, R. V. Tulchinskiy, 'Training in digital entrepreneurship as a basis for forming the intellectual capital of nation'. Information Technologies and Learning Tools, 2021 81(1), 210–221. https://doi.org/10.33407/itlt.v81i1.3899

[11] V. Koval, Y. Kremenetskaya, S. Markov, 'Promising Green Telecommunications Based on Hybrid Network Architecture'. 2019 International Conference on Information and Telecommunication Technologies and Radio Electronics (UkrMiCo), Odessa, Ukraine, 2019, pp. 1-4, doi: 10.1109/UkrMiCo47782.2019.9165525.

5

A New Decision-making Framework for Performance Evaluation of Industrial Robots

Bipradas Bairagi, Balaram Dey, and Goutam Bose

Department of Mechanical Engineering, Haldia Institute of Technology, India

Abstract

In today's extremely volatile global market scenario, industrial organizations are facing severe challenges to survive. In this circumstance, the management of industrial organizations is constantly searching for a way of making correct decisions in every aspect. Proper selection of robots for continuous and repetitive jobs in automotive manufacturing organizations is one of the most important and hard tasks for decision-makers. This hard task becomes harder and more complex when a decision is to be made with vague, imprecise, and ambiguous information in a fuzzy environment. This chapter aims to analyze the complexity of decision-making by exploring a new homogeneous group decision-making approach in robot selection, considering both tangible and intangible factors. A numerical example of a robot selection problem is illustrated. Assessment of performance rating of alternatives under subjective criteria as well as estimation of relative weights of selection criteria have been carried out based on the experience, opinion, and perception of the expert/decision-makers involved in the assessment process. The normalization process has been accomplished to restrict the magnitude and regulate the sense of the performance ratings of alternatives. The weights of the criteria and the normalized performance ratings have been integrated to calculate performance indices with benefit sense. The alternative having the highest degree of performance index has been selected as the best alternative. The results justify the applicability and validity of the proposed method.

Keywords: Robot selection, Fuzzy set theory, Tangible and intangible factor, Group decision making.

5.1 Introduction

Performance evaluation of industrial robots considering tangible and intangible factors is a very crucial decision-making procedure in the automotive manufacturing industry [1]. The selection criteria are classified as objective, subjective, and critical [1]. Objective criteria are those that are both measurable and tangible (quantitative) such as cost, weight, speed, distance, lifecycles measured in suitable units, etc. [1]. Subjective criteria are those that are intangible (qualitative) and neither measurable nor quantifiable [1]. Subjective criteria are associated with imprecision and vagueness [2]. Critical criteria are those that decide the requirement of further evaluation of data of an alternative. If critical criteria of an alternative are not satisfied, then the associated alternative is not considered for further evaluation. Political stability and social or communal situations are some examples of critical criteria [2, 3].

In the last few decades, extensive research on industrial robot selection has been executed. Some of the recent research works on robot selection have been cited in this section. Chatterjee *et al.* [4] used outranking and compromising ranking methods in the selection of robots for industrial purposes. Athawale and Chakraborty [5] employed a number of multiple-criteria decision-making approaches to make a comparative study of the industrial-based robot selection problem. Rao and Padmanabhan applied the matrix and diagraph method for the identification, comparison, and selection of robots for an industrial purpose [6]. Kumar and Prasad [7] proposed a new method for the selection of robots using theoretical and observed values for a specific application in industrial. Chu and Lin [8] applied a fuzzy-based technique of order preference by similarity to ideal solution (TOPSIS) method in the selection of robots under multi-criteria decision-making (MCDM) environment. Bhangale *et al.* [9] used an attribute-based method in the specification and selection of robots for specific functions.

Kahraman *et al.* [10] evaluated the robotic system for industrial application considering fuzzy multiple criteria. Karsak [11] used an integrated method based on regression analysis and quality function deployment for the selection of robots. Kumar and Garg [12] employed a distance-based method in the optimal selection of robots for performing specific functions. Tansel *et al.* [13] developed a decision support system for the purpose of robot selection in order to apply in managerial decision-making. Bairagi *et al.* [14] applied fuzzy multiple-criteria decision-making method for the evaluation and selection of robots in automated foundry operations. Liu *et al.* [15] developed an MCDM method using interval two-tuple-based linguistic techniques for robot evaluation, ranking, and selection. Parameshwaran *et al.* [16] proposed a combined fuzzy-based multi-criteria decision-making technique in the selection of robots taking both subjective and objective criteria into consideration. Bairagi *et al.* [17] developed a new MCDM technique for performance assessment and selection of material handling equipment. Joshi and Kumar [18] applied a TOPSIS-based MCDM tool with fuzzy choquet integral in robot selection. Rashid *et al.* [19] employed an integrated best−worst and evaluation based on distance from average solution (BW-EDAS), a multi-criteria decision-making method for the purpose of robot selection in industrial applications. Narayanamoorthy *et al.* [20] used an intuitionistic hesitant fuzzy VIseKriterijumska Optimizacija I Kompromisno Resenje (VIKOR) method based on entropy in industrial robot's evaluation and selection. Ali and Rashid [21] implemented the best−worst technique for the selection of robots. Shih [22] introduced an incremental analysis-based MCDM approach using group TOPSIS for the evaluation and selection of robots. Rashid *et al.* [23] applied the TOPSIS method with generalized interval-valued fuzziness for robot selection. Ghorabaee [24] developed an MCDM technique with interval-valued type-2 fuzzy sets for decision-making in robot selection.

The gap analysis of the above literature survey clearly shows that all the researchers have not addressed the issues of both tangible and intangible factors in robot selection problems. Even all papers did not consider the group decision-making process. Therefore, it is obvious that there is still the absolute necessity of further investigation for proper selection of the robots considering tangible and intangible factors with the objective of aiding managerial decision-makers.

The novelty of the proposed MCDM method is that it can consider the non-linearity features in assessing the performance index of industrial robots. The nonlinear characteristics are more real than linear ones in a complex uncertain decision-making environment. A real-life situation warrants the use of the nonlinearity as more logical and practical. Nonlinearity concept is a generic term, whereas linearity is a special case under it. Therefore, the proposed method has its unique advantage in accurate performance assessment over the other existing oversimplified, linear MCDM approaches used by the previous researchers.

The remaining part of the chapter has been arranged in the following manner. Section 5.2 is dedicated to describing the proposed algorithm. Section 5.3 cites a decision-making problem on robot performance evaluation and provides illustrative calculation. Section 5.4 is used for discussion. Lastly, Section 5.5 makes some essential concluding remarks.

5.2 Proposed Algorithm

This section introduces a new algorithm that can consider both tangible and intangible factors through the group decision-making process. The algorithm consists of the following 17 steps.

Step 1: Form a decision-making committee comprising the experts or decision-makers from different important sections of the organization. Denote the decision-makers as $[D_1, D_2, \ldots, D_p]$, where p is the number of the total decision-makers or experts.

Step 2: Select the selection criteria and denote the criteria as $[C_1, C_j, \ldots, C_n]$, where n is the number of the criteria.

Step 3: Execute an initial screening test for selecting the set of alternatives based on individual criterion cut-off value. Denote the alternatives as $A = \begin{bmatrix} A_1 & \ldots & A_i & \ldots & A_m \end{bmatrix}^T$, where m is the number of alternatives to be considered, and T stands for transpose matrix.

Step 4: Construct the decision matrix comprising performance ratings in terms of crisp numbers and linguistic variables.

$$B = \begin{matrix} & \begin{matrix} C_1 & \ldots & C_j & \ldots & C_n \end{matrix} \\ \begin{matrix} A_1 \\ \ldots \\ A_i \\ \ldots \\ A_m \end{matrix} & \begin{bmatrix} p_{11} & \ldots & p_{1j} & \ldots & p_{1n} \\ \ldots & \ldots & \ldots & \ldots & \ldots \\ p_{i1} & \ldots & p_{ij} & \ldots & p_{in} \\ \ldots & \ldots & \ldots & \ldots & \ldots \\ p_{m1} & \ldots & p_{mj} & \ldots & p_{mn} \end{bmatrix} \end{matrix}, \qquad (5.1)$$

where p_{ij} denotes the performance rating of the ith alternative with respect to the jth criterion.

Table 5.1 Linguistic variables, acronyms, and TFN for assessment of performance rating.

Linguistic variables	Acronyms	TFNs
Extremely low	EL	(0, 1, 3)
Low	L	(1, 3, 5)
Medium	M	(3, 5, 7)
High	H	(5, 7, 9)
Extremely high	EH	(7, 9, 10)

Step 5: Convert the linguistic variable into respective fuzzy number by using Table 5.1.

$$p_{ij} = \begin{cases} \tilde{p}_{ij}, & \text{if } j \in \text{intangible factor} \\ p_{ij}, & \text{if } j \in \text{tangible factor} \end{cases} \quad where \tilde{p}_{ij} = \left(p_{ij(1)}, p_{ij(2)}, p_{ij(3)}\right). \tag{5.2}$$

Step 6: Keep the crisp numbers unaltered and transform the fuzzy numbers into a crisp corresponding number using the following equation:

$$q_{ij} = \frac{1}{3} \sum_{k=1}^{3} p_{ij(k)}. \tag{5.3}$$

Step 7: Convert the entire performance ratings of the decision matrix in terms of crisp numbers only.

$$C = \begin{array}{c} \\ A_1 \\ \cdots \\ A_i \\ \cdots \\ A_m \end{array} \begin{array}{c} C_1 \quad \cdots \quad C_j \quad \cdots \quad C_n \\ \left[\begin{array}{ccccc} q_{11} & \cdots & q_{1j} & \cdots & q_{1n} \\ \cdots & \cdots & \cdots & \cdots & \cdots \\ q_{i1} & \cdots & q_{ij} & \cdots & q_{in} \\ \cdots & \cdots & \cdots & \cdots & \cdots \\ q_{m1} & \cdots & q_{mj} & \cdots & q_{mn} \end{array} \right] \end{array}, \tag{5.4}$$

where q_{ij} denotes the performance rating in crisp number ith alternative and jth criteria.

Step 8: Normalize the crisp decision matrix using the following normalization equation [25]:

$$r_{ij} = \frac{q_{ij} - (q_j)_{\min}}{(q_j)_{\max} - (q)_{\min}}, \tag{5.5}$$

where normalized rating is restricted by $0 \leq r_{ij} \leq 1$.

Step 9: Construct weight matrix with the knowledge, experience, and opinion of the experts involved in the decision-making process.

Table 5.2   Linguistic variables, acronyms, and TFNs for assessment of criteria weights.

Linguistic variables	Acronyms	TFNs
Equally important	EI	(1, 1, 2)
Inadequately important	II	(2, 3, 4)
Moderately important	MI	(4, 5, 6)
Strongly important	SI	(6, 7, 8)
Absolutely important	AI	(8, 9, 9)

$$
D = \begin{array}{c} \\ D_1 \\ \dots \\ D_i \\ \dots \\ D_p \end{array}
\begin{array}{c} C_1 \quad \dots \quad C_j \quad \dots \quad C_n \end{array}
\begin{bmatrix}
u_{11} & \dots & u_{1j} & \dots & u_{1n} \\
\dots & \dots & \dots & \dots & \dots \\
u_{i1} & \dots & u_{ij} & \dots & u_{in} \\
\dots & \dots & \dots & \dots & \dots \\
u_{p1} & \dots & u_{pj} & \dots & u_{pn}
\end{bmatrix}.
\tag{5.6}
$$

Here, u_{ij} represents the linguistic importance weight of criterion C_j assigned by the decision-maker D_i.

Step 10: Convert the linguistic variables of the weight matrix into fuzzy numbers (using Table 5.2) as follows:

$$
D = \begin{array}{c} \\ D_1 \\ \dots \\ D_i \\ \dots \\ D_p \end{array}
\begin{array}{c} C_1 \quad \dots \quad C_j \quad \dots \quad C_n \end{array}
\begin{bmatrix}
\tilde{v}_{11} & \dots & \tilde{v}_{1j} & \dots & \tilde{v}_{1n} \\
\dots & \dots & \dots & \dots & \dots \\
\tilde{v}_{i1} & \dots & \tilde{v}_{ij} & \dots & \tilde{v}_{in} \\
\dots & \dots & \dots & \dots & \dots \\
\tilde{v}_{p1} & \dots & \tilde{v}_{pj} & \dots & \tilde{v}_{pn}
\end{bmatrix}.
\tag{5.7}
$$

Here, $\tilde{v}_{ij}$ represents the fuzzy number representing weight of criterion C_j assigned by the decision-maker D_i.

Step 11: (a) Determine the mean weight in fuzzy number using the following formula:

$$
\tilde{v}_{j(\text{Mean})} = \left(\frac{1}{p}\sum_{i=1}^{p} v_{ij(L)}, \ \frac{1}{p}\sum_{i=1}^{p} v_{ij(M)}, \ \frac{1}{p}\sum_{i=1}^{p} v_{ij(U)} \right)
\tag{5.8}
$$

(b) Construct the mean weight matrix as follows:

$$
\begin{array}{c} C_1 \quad \dots \quad C_j \quad \dots \quad C_n \end{array} \\
\begin{bmatrix} \tilde{v}_{1(\text{Mean})} & \dots & \tilde{v}_{j\text{Mean}} & \dots & \tilde{v}_{n(\text{Mean})} \end{bmatrix}.
\tag{5.9}
$$

Step 12: Defuzzify the mean fuzzy weight($\bar{v}_j$) using the following equation:

$$\bar{v}_j = \frac{1}{3} \left(\frac{1}{p}\sum_{i=1}^{p} v_{ij(L)} + \frac{1}{p}\sum_{i=1}^{p} v_{ij(M)} + \frac{1}{p}\sum_{i=1}^{p} v_{ij(U)} \right). \quad (5.10)$$

L, *M*, and *U* denote lower, middle, and upper values, respectively.

Step 13: Evaluate the weight (w_j) for each criterion C_j using the following equation:

$$w_j = \frac{\bar{v}_j}{\sum_{i=1}^{n} \bar{v}_j}. \quad (5.11)$$

Step 14: Calculate the weighted normalized performance rating (x_{ij}) by applying the following equation:

$$x_{ij} = \text{EXP}(r_{ij} + w_j). \quad (5.12)$$

Step 15: Determine the performance score (PS_i) of the individual alternative A_i by the following equation:

$$PS_i = \sum_{j=1}^{n} \text{EXP}(r_{ij} + w_j). \quad (5.13)$$

Step 16: Measure the performance index (PI_i) for each individual alternative A_i by the following equation:

$$PI_i = \frac{\sum_{j=1}^{n} \text{EXP}(r_{ij} + w_j)}{\sum_{i=1}^{m} \sum_{j=1}^{n} \text{EXP}(r_{ij} + w_j)}. \quad (5.14)$$

Step 17: Arrange the alternatives in the decreasing order of their performance indices. Select the best alternative that has the highest performance index.

5.3 Illustrative Example

5.3.1 Problem definition

A manufacturing company desires to select the best robots for its new automotive factory to accomplish a specific task. For this purpose, a decision-making committee is formed with four members $(D_1, D_2, D_3,$ and $D_4)$ who are experts in different sections of the organization. The decision-maker D_1

Table 5.3   Decision matrix consisting of crisp and linguistic variables.

Robots	C_1 (+)	C_2 (+)	C_3 (−)	C_4 (−)	C_5 (+)	C_6 (+)
A_1	1.8	90	9500	0.45	L	EH
A_2	1.4	80	5500	0.35	M	H
A_3	0.8	70	4500	0.20	EH	EL
A_4	0.8	60	4000	0.15	H	M

belongs to the purchasing department; he is 50 years old and has 20 years of domain experience. The decision-maker D_2 belongs to the production department; he is 54 years old and has 24 years of domain experience. The decision-maker D_3 belongs to the finance department; he is 48 years old and has 16 years of domain experience. The decision-maker D_4 belongs to the management section; he is 58 years old and has 16 years of domain experience. The committee unanimously selects a set of six decision criteria, viz. velocity (C_1), load carrying capacity (C_2), costs (C_3), repeatability (C_4), vendor service quality (VSQ) (C_5), and programming flexibility (PF) (C_6). Four criteria (velocity, load carrying capacity, costs, and repeatability) are tangible (quantitative) criteria, whereas the remaining two criteria, service quality and programming flexibility, are intangible (qualitative) criteria. From the benefit and non-benefit angle of view, four criteria, viz. velocity, load carrying capacity, vendor's service quality, and programming flexibility are of the benefit category and the remaining two criteria, costs and repeatability, are of the non-benefit criteria. The decision-making committee chooses a set of four industrial robots (A_1, A_2, A_3, and A_4) through preliminary screening for further processes. The performance rating of the alternative robots under the tangible factors are obtained from different sources like the operating manuals, specifications, design handbook, and catalog. The performance ratings of the alternative robots under intangible factors as well as the importance weights of the criteria are assessed by the decision-making committee with their previous experience, knowledge, and valuable opinion in terms of linguistic variables. The linguistic variables for the assessment of alternatives for intangible factors and criteria weight are shown in Tables 5.1 and 5.2, respectively. The decision matrix is constructed by the decision-making committee in a combination of crisp numbers and linguistic variables as per eqn (5.1) and is depicted in Table 5.3. The proposed method has been applied below for finding the ranking order of the robot under consideration.

5.3.2 Calculation

The robot selection problems are summarized in the decision matrix and consist of both tangible and intangible factors in Table 5.3. The four homogeneous decision-makers are denoted by D_1, D_2, D_3, and D_4. The criteria have been designated by C_1, C_2, C_3, C_4, C_5, and C_6. The set of alternatives under consideration is denoted by A_1, A_2, A_3, and A_4. The linguistic variables expressing the performance ratings are converted into corresponding fizzy numbers using eqn (5.1). The decision matrix in terms of crisp is represented in Table 5.4. The conversion process of the linguistic variables into corresponding triangular fuzzy numbers is accomplished according to Table 5.1. The fuzzy numbers (shown in Table 5.4) of the decision matrix are converted into crisp numbers using eqn (5.3) and shown in Table 5.5. Eqn (5.4) has been used for this purpose. The maximum and minimum values of the crisp performance rating have been determined for each criterion and tabulated in the same column. The normalization process of the crisp performance ratings is accomplished to restrict the magnitude of the rating in the range of [0, 1] by employing eqn (5.5) and is depicted in Table 5.6. The weight matrix is formed by the decision-makers using eqn (5.6) for estimating the importance weight of the different criteria in terms of linguistic variables. The weight matrix is presented in Table 5.7. The linguistic variables of the weight matrix are transformed in terms of triangular fuzzy numbers using eqn (5.7) in the algorithm and shown in Table 5.8. Mean fuzzy weights and mean crisp weights of the criteria under consideration are computed using eqn (5.8). The mean weight matrix is constructed by using eqn (5.9). The mean fuzzy weight is defuzzified by using eqn (5.10). The weight for each criterion is calculated by using eqn (5.11). The estimated fuzzy weights, crisp weights, and normalized weights are furnished in Table 5.9. Weighted normalized decision matrix comprising weights of criteria and performance rating of alternative is calculated using eqn (5.12). The weighted normalized decision matrix is presented in Table 5.10. The performance score for each alternative robot is computed by using the prescribed eqn (5.13) and is shown in Table 5.11. The performance index (PI) for every alternative robot is measured by eqn (5.14) and put in Table 5.11. It is observed that alternatives A_1, A_2, A_3, and A_4 have the performance indices 0, 0.4548, 0.4904, and 1, respectively. Now, alternative robots are arranged in decreasing order of their performance indices as $A_4 > A_3 > A_2 > A_1$ and the corresponding rank is shown in Table 5.11. Therefore, A_4 is selected as the best robot having the highest performance index.

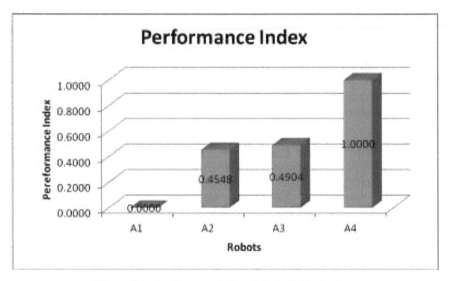

Figure 5.1   Performance index of the individual robot.

5.4 Discussions

In the current investigation, a decision-making problem on robot selection is considered. In the problem statement, there is a set of four feasible alternative robots, which are evaluated based on the six selection criteria under a committee comprising four homogeneous decision-makers. The proposed algorithm is applied for finding the appropriate decision. The calculation and the illustration show that robot A_4 attains the maximum performance index value (14.49). Therefore, robot A_4 is considered the most suitable one among all the available robots. Now, alternative robots are arranged in decreasing order of their performance indices as $A_4 > A_3 > A_2 > A_1$.

Table 5.4   Decision matrix in crisp and fizzy numbers.

Robots	C_1 (+)	C_2 (+)	C_3 (−)	C_4 (−)	C_5 (+)	C_6 (+)
A_1	1.8	90	9500	0.45	(1, 3, 5)	(1, 3, 5)
A_2	1.4	80	5500	0.35	(3, 5, 7)	(5, 7, 9)
A_3	0.8	70	4500	0.20	(7, 9,10)	(0, 1, 3)
A_4	0.8	60	4000	0.15	(5, 7, 9)	(7, 9,10)

Table 5.5 Decision matrix in crisp numbers with maximum and minimum values.

Robots	C_1 (+)	C_2 (+)	C_3 (−)	C_4 (−)	C_5 (+)	C_6 (+)
A_1	1.8	90	9500	0.45	3	3
A_2	1.4	80	5500	0.35	5	7
A_3	0.8	70	4500	0.2	8.66	1.33
A_4	0.8	60	4000	0.15	7	8.66
Max	1.8	90	9500	0.45	8.66	8.66
Min	0.8	60	4000	0.15	3	1.33

Table 5.6 Normalized performance rating.

Robots	C_1 (+)	C_2 (+)	C_3 (−)	C_4 (−)	C_5 (+)	C_6 (+)
A_1	1.00	1.00	1.00	1.00	0.00	0.23
A_2	0.60	0.67	0.27	0.67	0.35	0.77
A_3	0.00	0.33	0.09	0.17	1.00	0.00
A_4	0.00	0.00	0.00	0.00	0.71	1.00

Table 5.7 Weight matrix in terms of linguistic variables.

D_i	C_1	C_2	C_3	C_4	C_5	C_6
D_1	EI	MI	II	SI	EI	II
D_2	MI	AI	MI	SI	SI	MI
D_3	MI	SI	AI	MI	MI	AI
D_4	II	MI	MI	SI	MI	SI

Table 5.8 Weight matrix in terms of triangular fuzzy numbers.

D_i	C_1	C_2	C_3	C_4	C_5	C_6
D_1	(1, 1, 2)	(4, 5, 6)	(2, 3, 4)	(6, 7, 8)	(1, 1, 2)	(2, 3, 4)
D_2	(4, 5, 6)	(8, 9, 9)	(4, 5, 6)	(6, 7, 8)	(6, 7, 8)	(4, 5, 6)
D_3	(4, 5, 6)	(6, 7, 8)	(8, 9, 9)	(4, 5, 6)	(4, 5, 6)	(8, 9, 9)
D_4	(2, 3, 4)	(4, 5, 6)	(4, 5, 6)	(6, 7, 8)	(4, 5, 6)	(6, 7, 8)

Table 5.9 Fuzzy weight, crisp weight, and normalized weight of criteria.

Weight	C_1	C_2	C_3	C_4	C_5	C_6
Fuzzy weight	(2.75, 3.5, 4.5)	(5.5, 6.5, 7.25)	(4.5, 5.5, 6.25)	(5.5, 6.5, 7.5)	(3.75, 4.5, 6.5)	(5, 6, 6.75)
Crisp weight	3.58	6.42	5.42	6.5	4.92	5.92
Normalized weight	0.11	0.2	0.17	0.2	0.15	0.18

Table 5.10 Weighted normalized decision matrix.

Robots	C_1	C_2	C_3	C_4	C_5	C_6
A_1	3.03	3.32	1.19	1.22	1.16	1.50
A_2	2.03	2.38	2.45	1.70	1.65	2.59
A_3	1.12	1.70	2.94	2.81	3.16	1.20
A_4	1.12	1.22	3.22	3.32	2.36	3.25

Table 5.11   Performance score and ranking order.

Robots	PS_i	PI_i	Rank
A_1	11.43	0.0000	4
A_2	12.82	0.4548	3
A_3	12.93	0.4904	2
A_4	14.49	1.0000	1

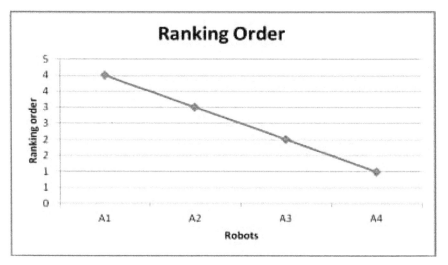

Figure 5.2   Ranking order of the individual robot.

5.5 Conclusion

Robot selection in the automotive industry is a very crucial decision-making process, and it is an imperative responsibility of managerial decision-makers of any organization. The decision-making process involves significant capital investment and time. The success of an automotive manufacturing company in many cases depends upon proper decision-making regarding the selection of proper robots that take a major role in future progress. Naturally, the decision-making committee considers multiple conflicting factors in a combination of tangible, intangible, benefit, non-benefit, or target-based criteria. In such a circumstance, the solution process of a complex decision-making problem deserves a new suitable MCDM model. The current investigation has developed and suggested a novel decision-making algorithm with 17 steps capable of considering tangible, intangible, benefit, and non-benefit criteria. The solution to the robot selection problem clearly shows that the proposed method is effective and useful for the selection of appropriate robots under tangible and intangible factors. Consideration of dependent criteria through a heterogeneous group decision-making process may be a direction of future research.

Acknowledgement

This work was supported by the Department of Mechanical Engineering, Haldia Institute of Technology, Haldia, West Bengal, India.

References

[1] B. Bairagi, B. Dey, B. Sarkar, S. Sanyal S, 'Selection of robotic systems in fuzzy multi criteria decision-making environment.' International Journal of Computational Systems Engineering 2 (2015) 32-42.

[2] A. Bhattacharya, B. Sarkar and S. K. Mukherjee, 'Material handling equipment selection under multi criteria decision making (MCDM) environment.' Industrial Engineering Journal 31 (6) (2002) 17–25.

[3] T-C. Chu and Y-C. Lin, 'A fuzzy TOPSIS method for robot selection.' International Journal of Advance Manufacturing Technology 21 (4) (2003) 284–290.

[4] P. Chatterjee, V. M. Athawale, S. Chakraborty, 'Selection of industrial robots using compromise ranking and outranking methods.' Robotics and Computer Integrated Manufacturing 26 (2010) 483–489.

[5] V. M. Athawale and S. Chakraborty, 'A comparative study on the ranking performance of some multi-criteria decision-making methods for industrial robot selection.' International Journal of Industrial Engineering Computations, 2(4) (2011) 831–850.

[6] R. V. Rao and K. K. Padmanabhan, 'Selection, identification and comparison of industrial robots using diagraph and matrix methods.' Robotics and Computer Integrated Manufacturing 22(4) (2006) 373–383.

[7] E. S. R. R. Kumar and J. S. R. Prasad, 'A novel approach of robot selection with the help of observed and theoretic- cal values for a given industrial application.' International Journal of Multidisciplinary and Current research 6 (2018) 532–535.2

[8] T. C. Chu and Y. C. Lin, 'A fuzzy TOPSIS method for robot selection.' International Journal of Advanced Manufacturing Technology (2003) 21(4) 284–29.

[9] P. P. Bhangale, V. P. Agrawal and S, K, Saha, Attribute based specifi cation, comparison and selection of a robot, Mechanism and Machine Theory 39(12) (2004) 1345–1366.

[10] C. Kahrama, S. Cevik, N. Y. Ates and M. Gulbay, 'Fuzzy multi-criteria evaluation of industrial robotic systems.' Computers & Industrial Engineering 52(4) (2007) 414–433.

[11] E. E. Karsak, 'Robot selection using an integrated approach based on quality function deployment and fuzzy regression.' International Journal of Production Research 46(3) (2008) 723–738.

[12] R. Kumar and R. K. Garg, 'Optimal selection of robots by using distance based approach method.' Robotics and Computer-Integrated Manufacturing 26(5) (2010) 500–506.

[13] Y. Tansel, M. Yurdakul and B. Dengiz, 'Development of a decision support system for robot selection.' Robot- ics and Computer-Integrated Manufacturing 29*(4)(2013) 142–157.

[14] B. Bairagi, B. Dey, B. Sarkar and S. Sanyal, 'Selection of robot for automated foundry operations using fuzzy multi-criteria decision making approaches.' International Journal of Management Science and Engineering Management (2014) 9(3) (2014) 221–232.

[15] H. Liu, M. Ren, J. Wu and Q. Lin, 'An interval 2-tuple linguistic MCDM method for robot evaluation and selec tion.' International Journal of Production Research 52(10) (2014) 2867–2880.

[16] R. Parameshwaran, S, P. Kumar and K. Saravanakumar, 'An inte-grated fuzzy MCDM based approach for robot selection considering objective and subjective criteria.' Applied Soft Computing 26 (2015) 31–41.

[17] B. Bairagi, B. Dey, B. Sarkar and S. K. Sanyal, 'A de novo multi-approaches multi-criteria decision making technique with an application in performance evaluation of material handling device.' Computers & Industrial Engineering 87 (2015) 267–282.

[18] D. Joshi and S. Kumar, 'Interval-valued intuitionistic hesitant fuzzy choquet integral based TOPSIS method for multi criteria group decision making.' European Journal of Operational Research 248(1) 2016183–191.

[19] T. Rashid, A. Ali and Y-M. Chu, 'Hybrid BW-EDAS MCDM method-ology for optimal industrial robot selection,' PLoS ONE 16(2) (2021) e0246738. https://doi.org/10.1371/journal. pone.0246738.

[20] S. Narayanamoorthy, S. Geetha, R. Rakkiyappan and YH. Joo, 'Interval-valued intuitionistic hesitant fuzzy entropy based VIKOR method for

industrial robot's selection.' Expert Systems with Applications 121 (2019) 28–37.

[21] A. Ali and T. Rashid, 'Best-worst method for robot selection.' Soft Computing, 26 (2020) 1–21.

[22] Hsu-Shih Shih, 'Incremental analysis for MCDM with an application to group TOPSIS.' European Journal of Operational Research 186 (2008) 720–734.

[23] T. Rashid, I. Beg and S. M. Husnine, 'Robot selection by using generalized interval-valued fuzzy numbers with TOPSIS.' Applied Soft Computing 21(2014) 462–468.

[24] M. K. Ghorabaee,'Developing an MCDM method for robot selection with interval type-2 fuzzy sets.' Robotics and Computer-Integrated Manufacturing 37 (2016) 221–232.

[25] L. Abdullah, W. Chan and A. Afshari, 'Application of PROMETHEE method for green supplier selection: a comparative result based on preference functions.' Journal of Industrial Engineering International 15 (2019) 271–285.

[26] P. Chatterjee and S. Chakraborty, 'Gear Material Selection using Complex Proportional Assessment and Additive Ratio Assessment-based Approaches: A Comparative Study.' International Journal of Materials Science and Engineering 1 (2013) 104-111.

6

Determination of Launch Time for a Multi-generational Product: A Fuzzy Perspective

Mohini Agarwal[1], Adarsh Anand[2], Chanchal[2], and Hitesh Kumar[2]

[1]Amity School of Business, Amity University, India
[2]Department of Operational Research, University of Delhi, India

Abstract

In the present market scenario, determining the launch time of a successive generational product is a critical decision. The timing decision depends upon whether the firm wants to push the product into the market before the competitors or wants to invest more time in perfecting the product design and development. The launch time is affected by different factors such as the cannibalization effect, promotional cost, the duration of promotion of the existing generational product, etc. An early introduction may hinder the current generation's sales, while a delay may result in the firm's loss of opportunity. In the past, many researchers have developed a model for the introduction time of successive generations. However, the majority of the work done is in a crisp environment. In the real market, parameters like cost and target sales of the previous generation are imprecise. So, the notion of fuzziness can be applied to cope with this ambiguity, which differentiates our study. Here, a fuzzy optimization problem has been formulated to capture this scenario, and the proposed work is illustrated with a numerical example.

Keywords: Fuzzy optimization, membership function, optimal introduction time, successive generation.

105

6.1 Introduction

Today, the world is even more competitive than it was before, which is analogous to any race competition [1]. To win, one must be faster than the competitors. In the competitive marketplace, companies need to innovate to hold the customers; failing in doing so can badly impact the company's performance, i.e., a significant drop in the sales figure can be seen. According to the well-known philosophy of Charles Darwin [11], "It is neither the strongest species nor the most intelligent species that survive, but the ones that are most responsive to changes in the surroundings survive." Principally, to survive, firms must come up with newer offerings for which marketers require new ideas, and, subsequently, more investment is required in research and development [3, 30]. By looking at the situation at hand and the acceptability of the existing product in the market, firms are coming up with strategies of making additions, feature enhancements, and improvements in the existing product with better performability, and at the same time being responsive to customers' demand. Marketers call the improved product with new additionalities as the generational product of the existing one [1, 2, 3, 5, 14].

When a company creates a new product, there are various kinds of expenses in it, from the idea of making the product to its adoption by the customer [12]. Because of this expense, it is a complex problem to decide when the company should introduce the new generation of the product into the market. In case the company introduces the new generation of its product into the marketplace in a short time frame, then the already existing product will not be able to earn even the cost incurred and, in this case, the company will suffer a loss. Just in case the company makes the new generation of its product available after a long time in the market, then in this situation, there can be an early entry from the competitors and the customers might not be ready to accept this generational product, and, in this way, the company will incur a loss [3]. Therefore, continuous improvement in terms of generational offerings and time to launch is a very important part of generating profit for the firm. Furthermore, the penetration of the product and its successive generations in the marketplace requires marketers' attention. According to firms, the main reason for the manufacturing of several generations of a product is to make the same idea more effective and get the maximum benefit from the help of that product. As the new generation of the product is more influential than before, customers will have confidence in the product; hence, they will remain attached to the product, and after the arrival of the new generation, they can easily accept the new generation [8].

The concept of diffusion of innovation has been explored by various researchers. In 1969, it was F. M. Bass whose attempt was focused on modeling the mixed influence model, which was suitable for a single generational product [6]. The work by Norton and Bass [18] brought a revolution in marketing by modeling the diffusion pattern of successive generational products. Subsequently, with the introduction of the new generation in the market, not everyone accepts the new generation immediately, which is why the demand for the previous generation continues even after the introduction of the newer generation [18] as illustrated in Figure 6.1.

Following Norton and Bass [18], other researchers developed extensions and variations to their model. Several researchers like Wilson and Norton [24] and Mahajan and Muller [17] have specifically focused on the case of one-time sales or the phase transition to elucidate the rule for the introduction of successive generations of the product either at now or never or now or at maturity. Padmanabhan and Bass [19] worked on the entry timing of the generational product and presented the optimal pricing policy. In the work by Purohit [20], the benefit of cost-effectiveness has been highlighted rather than the extension of loan for the substitutional product policy. In 1996, Mahajan and Muller proposed the mathematical structure for modeling the adopter's behavior and the skipping pattern in their adoption manner, whereas Cohen

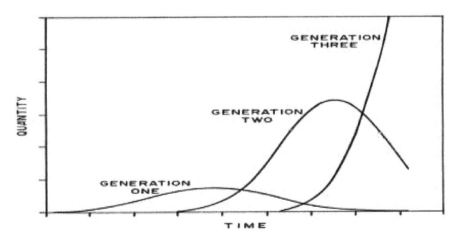

Figure 6.1 Series of technological generations.

et al. [10] assumed that products can only be sold during a fixed time window. Sohn and Ahn [22] demonstrated a cost−benefit analysis based on the Norton and Bass model [18] for computing the introduction time of a newer generation. It was in 2010 that a unique classification between the existence of switchers and substitutes was proposed by Jiang and Jain [14]. Kapur *et al.* [16] proposed a multi-generation model based on adoption, substitution, and repeat purchase to model the diffusion of televisions. Jiang and Jain [14] demonstrated that it is not necessary to introduce the newer generation only at "now" or "at maturity"; it can lie in between them. Aggrawal *et al.* [3] have worked on generational products and determined the launch time of the second generation of the product by framing an optimal launch time policy. Thus, the launch time determination has become a serious concern for any firm as too early or delayed has its own merits and demerits [3].

There can be a few instances in which after coming up with the newer generation, the potential adopter's base will reduce drastically, and any delay in the introduction time creates negative thinking because of the competitor's presence in the market. Consequently, several researchers have worked on finding out the introduction period of the next generation and the crisp method has been used in most of the studies. However, in real life, the input parameters for the determination of launch time are often imprecise or fuzzy. Zimmermann [26] first introduced a fuzzy set theory into conventional linear programming problems with a fuzzy goal and fuzzy constraints. Using the Bellman and Zadeh [7] fuzzy decision-making method, Zimmermann[26] proposed the equivalent linear programming problem based on membership function. Practitioners like Jha and Aggarwal [13] have worked in a fuzzy environment to develop optimal advertising media allocation problems and the work by Anand *et al.* [4] was based on framing optimal scheduling policy for multi-upgraded software systems under a fuzzy environment. The current literature still lacks the determination of the introduction time obtained by employing the fuzzy method. Therefore, here, a fuzzy linear programming model for the determination of the time to introduce the new generation has been developed. Here, the framework with two-generation has been developed, out of which the first generation is already there in the market while the company wants to determine the introduction time of the second generation under fuzzy constraints on cost and on the cumulative sales of the earlier generation to attain the desired goal.

This chapter is divided into sections and organized as follows. Section 6.1 comprises of introduction along with the relevant literature review. In Section 6.2, the basic building blocks where the relationship between sales over time with the help of the Bass model [6] and the cost modeling framework as given by Aggrawal *et al.* [3, 29] has been discussed. Following it, in Section 6.3, an optimization problem to determine the launch time for the second generation under a fuzzy environment has been formulated. In Section 6.4, the numerical illustration has been discussed followed by managerial implications and conclusion in Sections 6.5 and 6.6, respectively. Lastly, a list of references has been given.

6.2 Building Block

In line with the work of Aggrawal *et al.* [3], the notion here lies in understanding the diffusion process and the formulation of the cost modeling structure. Before the determination of introduction time under an uncertain environment, briefly, the model assumptions are as follows:

- Here, the case of two successive generations of the product has been considered, i.e., an earlier generation already exists, and marketers wish to know the launch time of its succeeding version [28].
- The diffusion pattern of the existing generation evolves.
- The organization is cost-oriented, i.e., the marketer wishes to minimize the cost of production as well as the promotional cost over the period.
- The promotion of the existing generation of the product is carried out before the launch of successive generations of the product.
- The present cost structure is independent of the sales growth pattern of the existing generation and not of the upcoming generational product.

6.2.1 Innovation diffusion model [6]

Being the most widely used innovation diffusion model, the work by F. M. Bass [6] is well suited for a single generation of the product as it considers the finite number of potential adopters who, with time, purchase the product under two influences, viz. the direct influence (information regarding the product is directly received) and indirect influence (knowledge about the product is received from existing adopters). The model as given by F. M. Bass [6] will be used for the determination of introduction time and it can be given as follows:

$$n\left(t\right) = \frac{dN\left(t\right)}{dt} = \left(p + q\frac{N\left(t\right)}{m}\right)\left(m - N\left(t\right)\right), \qquad (6.1)$$

where "p" and "q" represent the coefficient of direct and indirect influences;"m" represents the number of prospective buyers. Eqn (6.1) gives the prospective adopters of the product at any time point$N$ $(t = 0) = 0$, N $(t \to \infty) = m$. Using the above boundary conditions and on solving eqn (6.1), the cumulative number of adopters of the first generation before the launch of its succeeding generations can be given as follows:

$$N\left(t\right) = m \left(\frac{1 - e^{-(p+q)t}}{1 + \left(\frac{q}{p}\right) e^{-(p+q)t}} \right) \tag{6.2}$$

$$n\left(t\right) = m \frac{(p+q)^2}{p} \left(\frac{1 - e^{-(p+q)t}}{\left(1 + \left(\frac{q}{p}\right) e^{-(p+q)t}\right)^2} \right). \tag{6.3}$$

Eqn (6.2) and (6.3) are important for marketers as it provides the cumulative as well as the non-cumulative number of adopters of the product. The sales pattern of the same with the inflection point is given in Figure 6.2; it

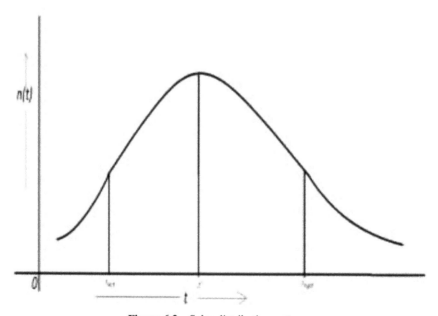

Figure 6.2   Sales distribution pattern.

highlights the peak sales time and peak sales as well, which can be obtained by solving eqn (6.3) and given as follows:

$$t^* = \frac{1}{(p+q)} \ln\left(\frac{q}{p}\right) \tag{6.4}$$

and

$$n(t^*) = \frac{m}{4q}(p+q)^2. \tag{6.5}$$

Eqn (6.4) and (6.5) enable the marketers to determine the time for the peak sales t^* and the magnitude $(n(t^*))$. The above equations will be used in the determination of introduction time.

6.2.2 The cost model [3]

To understand the introduction of advanced generations based on the existing generation, the cost of production and promotion for previous generation products has been kept in mind. The cost structure as given by Aggrawal *et al.* [3] is well suited under monopolistic market conditions and assumed that each potential buyer will only purchase one product. By making use of this model, the focus is to determine the optimal launch time in a fuzzy environment under cost as well as sales constraints.

In the model, Aggrawal *et al.* [3] have assumed that $(C_1 \leq C_2)$, i.e., after the introduction of the second generation, the production cost of the first generation will be high as compared to before its introduction due to the shift in market paradigm. The prospective adopters will adopt the recent generation as compared to its previous generation; thus, the firm's focus will be on producing and promoting the newer generation more, leading to an increase in the cost of its previous version. Under the assumptions [3], the cost structure $C(T)$ is given by [12]

$$C(T) = C_1 N(T) + C_2(m - N(T)) + C_3 T. \tag{6.6}$$

In eqn (6.6), C_1 is the cost associated with the production of the older generation before the introduction of the newer generation $(t \leq T)$; C_2 is the cost of production of the older generation after the introduction of the newer generation in the marketplace (i.e., $t > T$), and C_3 is the promotional cost per unit time for the first-generation product (i.e., $t \leq T$). To determine the optimal introduction time T^*, the mathematical problem under the cost constraint can be written as

$$\text{Min } C(T) = C_1 N(T) + C_2(m - N(T)) + C_3 T$$
subject to

(P-1)

$$C(T) \leq C_B$$
$$N(T) \geq N_0.$$

Here, C_B is the amount as budget and N_0 is target sale. The company aspires to determine the introduction time with the objective of cost minimization with constraints on production and promotional costs and target sales.

6.3 The Fuzzy Optimization Problem

The optimization problem (P-1) under a fuzzy environment can be formulated as follows:

$$\text{Min } C(T) = C_1 N(T) + C_2(m - N(T)) + C_3 T$$
subject to

(P-2)

$$C(T) \tilde{\leq} C_B$$
$$N(T) \tilde{\geq} N_0$$

where the inequality in the cost and target sales for the first-generation constraint is not precisely known. The management may be ready to bear the slight increase/decrease in investment and target sales. The inequality signs in eqn (P-2) represent the fuzzy less than or equal to and greater than or equal to and have associated linguistic meaning, i.e., it can be essentially less than or equal to and greater than or equal to. For solving fuzzy mathematical problems, Zimmermann's [25] approach has been used. In this methodology, first, the objective function of the fuzzifier minimization is perceived as a limitation level and appropriate membership functions are characterized for fuzzy disparities; then Bellman and Zadeh's [7] approach has been applied to distinguish the fuzzy result, which outcomes in a crisp mathematical programming problem algorithm.

Find T
subject to $C(T) < C_B$

(P-3)

$$N(T) > N_0$$
$$T > 0.$$

The authors define the membership functions $\mu_1(T)$ and $\mu_2(T)$ and the fuzzy inequality in the problem (P-2) [9].

$$\mu_1 = \begin{cases} 1 & \text{if } N(T) \geq N_0 \\ \frac{N(T) - N^*}{N_0 - N^*} & \text{if } N^* < N(t) \leq N_0 \\ 0 & \text{if } N(T) \leq N^* \end{cases}$$

$$\mu_2 = \begin{cases} 1 & \text{if } C(T) \le C_0 \\ \frac{C_B - C(T)}{C_B - C_0} & \text{if } C_0 \le C(T) \le C_B \\ 0 & \text{if } C(T) \ge C_B \end{cases}$$

where N^* and C_0 are respective tolerance limits on target sales level and the investment the company is willing to make, wherein C_0 is the minimum investment for the existing generation production and promotion and C_B represents the maximum available budget to cater to the market before the introduction of advanced generation in the marketplace. Bellman and Zedeh's [7] theory has been utilized to characterize the fuzzy decision to solve the fuzzy system of inequality related to the problem. The simplified crisp optimization problem is

Maximize α

$\mu_i(T) \ge \alpha$ for $i = 1, 2$

$1 \ge \alpha \ge 0$ (P-4)

$T \ge 0.$

The problem (P-4) after incorporating values can be solved by a mathematical programming approach using LINGO or any other software package.

6.4 Numerical Illustration

In this section, the above introduction of the new generation policy is discussed by using DRAM's data set [23]. The authors have specifically considered the sales of 4k DRAM having 12 years of observations for estimation purposes. The parameters are obtained by estimating eqn (6.2) on a statistical software package widely known as SAS [21]. Following are the estimated values: $m = 318.56, p = 0.0189,$ and $q = 0.921$. Further, let us assume the cost parameters as $C_1 = \$70, C_2 = \$90,$ and $C_3 = \$40$. Also, the company is willing to invest $C_0 = \$19500$ and target sales for the existing version $N_0 = 240$. The tolerance limits on investment and target sales are $C_B = \$25,000$ and $N^* = 160$ (assumed as base values on the market management based on the experience relating to the introduction of the new generation). The cost function given by eqn (6.6) under the assumed set of values can be graphically represented by Figure 6.3, whereas the sales growth function given by eqn (6.2) is given in Figure 6.4. The membership functions are represented graphically in Figure 6.5.

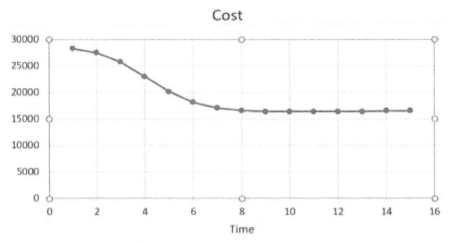

Figure 6.3   The cost framework for existing generation.

Figure 6.4   Sales growth pattern.

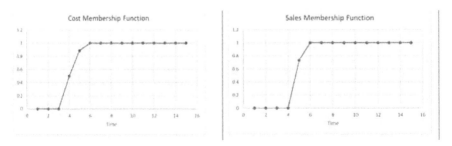

Figure 6.5   Cost and sales membership function.

The membership function for the fuzzy cost and target sales constraint is given as

$$\mu_1(T) = \frac{25,000 - \left(\begin{array}{c}70*318.56\left(\frac{1-e^{-(0.9399)T}}{1+48.73e^{-(0.9399)T}}\right) + \\ 80\left(318.56 - 318.56\frac{(1-e^{-(0.9399)T})}{(1+48.73e^{-(0.9399)T})}\right) + 40T\end{array}\right)}{25,000 - 19,500},$$

where $19,500 \le C(T) \le 25,000$

$$\mu_2(T) = \frac{\left(318.56\frac{(1-e^{-(0.9399)T})}{(1+48.73e^{-(0.9399)T})}\right) - 160}{240 - 160} \quad \text{where } 160 \le N(T) \le 240.$$

Using the above values and constraints, the solution to the problem (P-4) using the fuzzy optimization technique is obtained as

Maximize α
Subject to

$$\left[C_B - \left\{\begin{array}{c}C_1 m\left(\frac{1-e^{-(p+q)T}}{1+\frac{q}{p}e^{-(p+q)T}}\right) + \\ C_2\left(m - m\frac{1-e^{-(p+q)T}}{1+\frac{q}{p}e^{-(p+q)T}}\right) + C_3 T\end{array}\right\}\right] \ge (C_B - C_0)\,\alpha$$

$$\left[m\left(\frac{1-e^{-(p+q)T}}{1+\frac{q}{p}e^{-(p+q)T}}\right) - N^*\right] \ge (N_0 - N^*)\,\alpha \qquad \text{(P-5)}$$

$1 \ge \alpha \ge 0$
$T \ge 0.$

By putting the value of assumed constants, eqn (P-5) reduces in the following form:

Maximize α
Subject to

$$\left[25,000 - \left\{\begin{array}{c}\left\{70*318.56\left(\frac{1-e^{-(0.9399)T}}{1+48.73e^{-(0.9399)T}}\right)\right\} \\ +\left\{80\left(318.56 - 318.56\frac{(1-e^{-(0.9399)T})}{(1+48.73e^{-(0.9399)T})}\right)\right\} \\ +40T\end{array}\right\}\right]$$

$$\ge 55,00\alpha$$

$$\left[\left(318.56\frac{(1-e^{-(0.9399)T})}{(1+48.73e^{-(0.9399)T})}\right) - 160\right] \ge 80\alpha \qquad \text{(P-6)}$$

$0 \le \alpha \le 1, \quad T \ge 0.$

On solving eqn (P-6), the introduction time of successive versions $T^* = 5.35, \alpha^* = 1$ has been obtained. The cost $C(T^*) = 19,284.61$

and $N(T^*) = 203.39$. The above results depict the launch time of the next version, which should be at 5.35 years under the fuzzy environment.

6.5 Managerial Implications

During the initial launch of the product, the initial adoption of the product is slow. With promotional activities, there is a gradual increase in its adoption rate. Marketers do everything to promote the product, which will increase the sales of the product and once it reaches the saturation level, there is the need for new technological/innovative products, which will compete and replace the existing product, thereby making it necessary from a managerial perspective to understand and determine the time at which the innovative/successive version of the product is to be launched in the marketplace. However, the early or late introduction has its own merits and demerits, but as a decision-maker, it is wise to take consumer expectations and associated determinants into consideration. The market situations keep varying; thus, as a decision-maker, the focus should be on such an optimization model that can accommodate these variabilities in different market parameters.

Here, the proposed optimization model is formulated under a fuzzy environment to have suitable flexibility in terms of budget and target sales levels as planned by marketers. The optimization framework will help in managerial decisions of successive generation product launches, which will minimize the production and promotion cost and meet the target sales level under uncertain situations. Decisions about launch time should be based on several factors, viz. the cannibalization effect, the attributes that are indicators of customer expectations, the product development, promotion cost, and duration of promotion of the existing generational product. From the managerial perspective, the focus should be on all the contrasting attributes that can be evaluated with the help of the proposed framework; being a fuzzy problem, it provides flexibility and helps decision-makers in launch time determination.

6.6 Conclusion

Managing the life cycle for generational products poses a great challenge to marketers, and the optimal introduction timings for the generational product are one of the toughest decisions that a manager must take. The fundamental thought behind it is to catch the advantages of advances made by the company's R&D team about creating new highlights, improving the product configuration, creating innovative products, and so on over the long run. The decision for optimal introduction timings of the generational product depends upon whether organizations should put additional time into improving the product configuration and its development or launch the product before the competitive product comes into the market. The optimization model discussed in the chapter is dynamic to accommodate the globalized nature and the cost model is explicitly based on the demand of the existing generation. The current work helps in planning for the succeeding version of the product based on the existing version's performance in the market and the investment the company is willing to make. The optimization policy as formulated in this chapter helps in computing the optimal introduction time to launch the newer version keeping in mind that the earlier version has performed well, i.e., introduction time can assist in planning the strategies that can help in terms of revenue generation and, at the same time, help in establishing the competitive edge.

Acknowledgement

The authors would sincerely like to thank the editor and the reviewers for their valuable comments and suggestions, which helped to significantly improve the quality of the chapter.

References

[1] M. Agarwal, D. Aggrawal, A. Anand, and O. Singh, "Modeling multi-generation innovation adoption based on conjoint effect of awareness process," International Journal of Mathematical, Engineering and Management Sciences, vol. 2, no. 2, pp. 74-84, 2017.

[2] D. Aggrawal, A. Anand, O. Singh, and P. K. Kapur, "Modelling successive generations for products- in-use and number of products sold in the market," International Journal of Operational Research, vol. 24, no. 2, pp. 228–244, 2015.

[3] D. Aggrawal, O. Singh, A. Anand, and M., Agarwal, "Optimal introduction timing policy for a successive generational product," International Journal of Technology Diffusion, vol. 5, no. 1, pp. 1–16,2014.

[4] A. Anand, S. Das, M. Agarwal, and V. S. S. Yadavalli, "Optimal Scheduling Policy for a Multi-upgraded Software System under Fuzzy Environment," Journal of Mathematical and Fundamental Sciences, vol. 51, no. 3, pp. 278-293, 2019.

[5] A. Anand, O. Singh, D. Aggrawal, and J. Singh, "An interactive approach to determine optimal launch time of successive generational product," International Journal of Technology Marketing, vol. 9, no. 4, pp. 392–407, 2014.

[6] F. Bass, "A new product growth for model consumer durables," Management Science, vol. 15, no. 5, pp. 215–227, 1969.

[7] R. E. Bellman, and L. A. Zadeh, "Decision-making in a fuzzy environment," Management Science, vol. 17, no. 4, pp. B141-B164, 1970.

[8] U. Chanda, "The dynamic price-quality decision model for two successive generations of technology innovation," International Journal of Innovation and Technology Management, vol. 8, no. 4, pp. 635–660, 2011.

[9] S. Chaube and S. B. Singh, "Fuzzy reliability theory based on membership function," International Journal of Mathematical, Engineering and Management Sciences, vol. 1, no. 1, pp. 34-40. 2016

[10] M. A. Cohen, J. Eliashberg, J., and T. H. Ho, "New product development: The performance and time-to-market trade-off," Management Science, vol. 42, no. 2, pp. 173–186, 1996.

[11] C. Darwin, The origin of species (pp. 95-96). New York: PF Collier & Son.1909.

[12] A. H. S. Garmabaki, P. K. Kapur, J. N. P. Singh, and R. Sanger, "The optimal time of new generation product in the market," Communications in Dependability and Quality Management-An International Journal, vol. 5, no. 1, pp. 123–137, 2012.

[13] P. C. Jha and R. Aggarwal, "Optimal advertising media allocation under fuzzy environment for a multi-product segmented market," Turkish Journal of fuzzy systems, vol. 3, no. 1, pp. 45-64, 2012.

[14] Z. R. Jiang and D. C. Jain, "Optimal marketing entry timing for successive product/service generations," Working Paper, 2012.

[15] Z. R. Jiang and D. C. Jain, "A generalized Norton-Bass model for multigeneration diffusion," Management Science, vol. 58, no. 10, pp. 1887–1897, 2012.

[16] P. K. Kapur, U. Chanda, A. Tandon, and S. Anand, "Innovation diffusion of successive generations of high technology products," in Proceedings of the 2nd International Conference on Reliability, Safety and Hazard (ICRESH), Mumbai, India, 2010. pp. 505–510.

[17] V. Mahajan and E. Muller, "Timing, diffusion, and substitution of successive generations of technological innovations: The IBM mainframe

case," Technological Forecasting and Social Change, vol. 51, no. 2, pp. 109–132, 1996.

[18] J. A. Norton and F. M. Bass, "A diffusion theory model of adoption and substitution for successive generations of high-technology products," Management Science, vol. 32, no. 9, pp. 1069–1086, 1987.

[19] V. Padmanabhan and F. M. Bass, "Optimal pricing of successive generations of product advances," International Journal of Research in Marketing, vol. 10, no. 2, pp. 185-207, 1993.

[20] D. Purohit, "What should you do when your competitors send in the clones?" Marketing Science, vol. 13, no. 4, pp. 329–411, 1994.

[21] SAS/ETS 9.1 User's Guide, SAS Publishing, SAS Institute Inc., Cary, NC, USA, 37-45, 2004.

[22] S. Y. Sohn and B. J. Ahn, "Multigeneration diffusion model for economic assessment of new technology," Technological Forecasting and Social Change, vol. 70, no. 3, pp. 251–264, 2003.

[23] N. M. Victor and J. Ausubel, "DRAMs as model organisms for study of technological evolution" Technological Forecasting and Social Change, vol. 69, no. 3, pp. 243-262, 2002.

[24] L. O. Wilson and J. A. Norton, "Optimal entry timing for a product line extension," Marketing Science, vol. 8, no. 1, pp. 1–17, 1989.

[25] H. J. Zimmermann, Fuzzy set theory—and its applications. Springer Science & Business Media, 2011.

[26] H. J. Zimmermann, "Description and optimization of fuzzy systems," International journal of general System, vol. 2, no. 1, pp. 209-215, 1975.

[27] N. Meade and T. Islam, "Modelling and forecasting national introduction times for successive generations of mobile telephony," Telecommunications Policy, vol. 45, no. 3, pp. 102088, 2021.

[28] A Anand, M Agarwal, D Aggrawal, L Hughes, P Maroufkhani, Y K Dwivedi, "Successive generation introduction time for high technological products: an analysis based on different multi-attribute utility functions". Environment, Development and Sustainability. Apr 28:1-8., 2022.

[29] D Aggrawal, A Anand, G Bansal, G H Davies, P Maroufkhani, Y K Dwivedi. "Modelling product lines diffusion: a framework incorporating competitive brands for sustainable innovations." Operations Management Research. May 11:1-3., 2022.

[30] A Anand, D Aggrawal, M Agarwal. Market assessment with OR applications. CRC Press; Dec 6., 2019.

7

Securing the Key of Improved Playfair Cipher using the Diffie–Hellman Algorithm

Raksha Verma[1], Riya Verma[2], and Adarsh Anand[3]

[1]Shaheed Rajguru College of Applied Sciences for Women, University of Delhi, India
[2]Miranda House, University of Delhi, India
[3]Department of Operational Research, University of Delhi, India

Abstract

Cryptography is the science of converting any secret information into indecipherable text to make it secure and immune to attacks. It consists of various techniques that allow only the sender and the receiver to understand the information meaningfully. The Wheatstone Playfair Square is a symmetric substitution encryption technique that is traditionally weak and easy to break. The main motive of this chapter is to make the traditional Playfair cipher more secure and attack-proof by securing the keyword of the Playfair cipher using the Diffie–Hellman algorithm and by changing the techniques of the Playfair cipher to an extended/improved version. The chapter introduces a new and extended version of the Playfair cipher, which has the role of a private key in encryption and decryption. The chapter has proposed a new algorithm using fuzzy logic with the aim to make communication more secure. The process will have two stages, wherein the first stage, the Diffie–Hellman algorithm will be used to exchange the secret key between the sender and the receiver. Furthermore, the secret key exchanged by the Diffie–Hellman algorithm will be used to find the private key for the extended Playfair cipher. In the second stage, the traditional Playfair cipher technique will be modified by changing the way of doing encryption and decryption, as now the secret key will determine the movement of alphabets in the Playfair square. Finally, the advantages of this new combination will be analyzed.

Keywords: coordinate address, decryption, Diffie–Hellman algorithm, encryption, fuzzy logic, monarchy, public key, private key, Playfair cipher, secret key, soft computing.

7.1 Introduction

Soft computing is the partnership of different methods that in one way or another conform to its guiding principle [1]. One of the important fields of soft computing is the concept of fuzzy logic. Combining the concept of fuzzy logic with the pre-defined algorithms of cryptography can form a completely new and secure technique to protect data. The secret sharing concept using fuzzy logic is an enhanced method of encoding for more security [2]. There have been many papers that introduce the new cryptography algorithm with fuzzy logic with low process time and high-security logic [3].

7.1.1 Introduction about cryptography and substitution techniques

Cryptography is associated with the process of converting plaintext into an unintelligible text, in order to secure meaningful information. It is the science of secret writing with the goal of hiding the meaning of a message [4]. It protects against browsing by making the information incomprehensible. It can supplement access controls and is especially useful for protecting data on tapes and discs, which, if stolen, can no longer be protected by the system [5].

The two basic building blocks of all encryption techniques are substitution and transposition. Substitution techniques play a vital role when it comes to different techniques in cryptography. A substitution technique is one in which the letters of plaintext are replaced by other letters or by numbers or symbols [6]. There are a number of cipher techniques that fall under the category of substitution. Some of the popular techniques are as follows:

1. Caesar cipher
2. Hill cipher
3. Playfair cipher

Our chapter deals with the modification and extension of the Playfair cipher technique, which is traditionally a symmetric substitution technique.

7.1.2 Introduction of the traditional playfair cipher

The Wheatstone–Playfair cipher is a manual substitution symmetric encryption technique and is the best-known multiple-letter encryption cipher. The technique was invented in 1854 by Charles Wheatstone but bears the name of Lord Playfair for promoting its use. The Playfair cipher algorithm is based on the use of a 5×5 matrix of alphabets constructed using a keyword. The keyword is known as "monarchy." The matrix is constructed by filling the alphabets of keywords from left to right and top to bottom, filling the remaining matrix with the remaining alphabets in the alphabetic order. Since the matrix can contain only 25 alphabets, it is needed to replace one alphabet with some other; for example, normally, in Playfair cipher, the alphabet "J" is replaced by the alphabet "I." After decryption of the message, the user needs to determine whether the alphabet "I" or "J" is making sense in the decrypted message.

For the process of encryption, a user needs to break the message into digraphs (pairs of two alphabets) and if the total number of alphabets is "odd" in number, then one can add a default alphabet "Z" to make the last pair complete. For example, the text "NAMASTE INDIA" becomes "NA, MA, ST, EI, ND, IA." After this, each digraph needs to be individually encrypted using the Playfair matrix by following the given rules.

1. If both the alphabets of a digraph are in the same row of the Playfair matrix, then replace the alphabets with their adjacent right alphabets, respectively (going back to the leftmost if at the rightmost).
2. If both the alphabets of a digraph are in the same column of the Playfair matrix, then replace the alphabets with their adjacent below alphabets, respectively (going back to uppermost if at the lowermost).
3. If both the alphabets of a digraph are not on the same row or the same column, then form a rectangle with the two alphabets as corners and take the alphabets on the horizontal opposite corner of the rectangle.

Similarly, for the process of decryption, one can follow the same rules in the opposite manner.
Example: Keyword – "CAR"
Plaintext – "HEAL"

C	A	R	B	D
E	F	G	H	I
K	L	M	N	O
P	Q	S	T	U
V	W	X	Y	Z

Using the process of encryption of the Playfair cipher, the plaintext is divided into digraphs: "HE, AL."

Consider the digraph "HE." Clearly, both the letters are in the same row. So, the letter "H" is encrypted to the letter "I" and the letter "E" is encrypted to the letter "F," respectively.

Consider the digraph "AL." Clearly, both the letters are in the same column. So, the letter "A" is encrypted to the letter "F" and the letter "L" is encrypted to the letter "Q," respectively.

Combining both the encrypted digraphs, the ciphertext is "IFFQ."

7.2 Works and Modifications Done on the Playfair Cipher Till Now

There have been many modifications in the traditional Playfair cipher, namely by introducing a rectangular 10×9 matrix [7] or an 8×8 matrix [8] instead of 5×5 to overcome the shortfalls of the traditional Playfair cipher, by presenting a Playfair CBC encryption mechanism to make the cryptanalysis complex [9], by enhancing the existing algorithm with a 6×6 matrix [10], by modifying traditional Playfair cipher by using a 7×7 matrix with a matrix randomization algorithm to extend the data holding capability and security at the same time [11], by using a 6×6 matrix instead of a 5×5 one, which includes elements from the digit set and the special character underscore [12], by using Playfair and Caesar ciphers in substitution techniques [13], by including two symbols "*" and "#" in a 7×4 matrix [14], or by exchanging the key of Playfair cipher using the RSA algorithm [15] and many more.

7.3 What Makes This Paper Unique?

This chapter introduces a new extended version of Playfair cipher, which has the role of a private key in encryption and decryption. Also, knowing the fact that the Diffie−Hellman algorithm is not considered vulnerable to attack, the chapter is using the Diffie–Hellman algorithm to exchange the secret key for the Playfair keyword (monarchy) between the users more safely. We have also changed the encryption and decryption algorithms of the traditional Playfair cipher by introducing the role of a secret key for the movement of alphabets in the Playfair matrix.

7.4 Building Blocks of the Proposed Work

7.4.1 The extended playfair cipher

The process of generating the Playfair square of the 5 × 5 matrix and pairing the alphabets of the plaintext into digraphs remains the same as in the traditional Playfair cipher. The matrix is constructed by filling the alphabets of keywords from left to right and top to bottom, filling the remaining matrix with the remaining alphabets in the alphabetic order. Since the matrix can contain only 25 alphabets, it is needed to replace one alphabet with some other; for example, normally in the Playfair cipher, the alphabet "J" is replaced by the alphabet "I." After decryption of the message, the user needs to determine whether the alphabet "I" or "J" is making sense in the decrypted message. The user needs to break the message into digraphs (pairs of two alphabets) and if the total number of alphabets is "odd" in number, then one can add a default alphabet "Z" to make the last pair complete. For example, the text "NAMASTE INDIA" becomes "NA, MA, ST, EI, ND, IA."

Now, in the extended Playfair cipher, we will make use of a secret/private key. The private key can be any numerical value; let the private key number be denoted by "P." Since we are using a 5 × 5 matrix and the private key has to determine the movement of the alphabets in the Playfair square at the time of encryption and decryption, the value of the private key needs to be used as

$$P' = (p \text{ modulo } 5) + 1.$$

The value of P' will be used in the extended Playfair cipher.

7.4.1.1 Algorithm to encrypt the plaintext
Each digraph needs to be individually encrypted using the following rules.

1. **If both the alphabets of a digraph are in the same row:**
   Take the alphabet at the right (P')th place of each alphabet of the digraph in the Playfair matrix (going back to the leftmost if at the rightmost respectively).

2. **If both the alphabets of a digraph are in the same column:**
Take the alphabet at the (P')th place below each alphabet of the digraph in the Playfair matrix (going back to the topmost if at the very bottom respectively).

3. **If the alphabets of the digraph are neither in the same column nor the same row:**
Then if (x, y) denotes the coordinate address of one of the alphabets of the digraph in the Playfair matrix, the coordinate address of the encrypted alphabet in the Playfair matrix, say (a, b), equals to

$$([(x + P')\text{modulo } 5] + 1, [(y + P')\text{modulo } 5] + 1)$$

7.4.1.2 Algorithm to decrypt the ciphertext

The process of splitting the ciphertext into pairs of two alphabets called "digraphs" remains the same as in the encryption algorithm. Here, the number of alphabets in the ciphertext cannot be odd in number (since if "odd" during encryption, the user must have added a default alphabet "Z" to complete the pair).

Each digraph needs to be individually decrypted using the following rules.

1. **If both the alphabets of a digraph are in the same row:**
Take the alphabet at the left (P')th place of each alphabet of the digraph in the Playfair matrix (going back to the rightmost if at the leftmost respectively).

2. **If both the alphabets of a digraph are in the same column:**
Take the alphabet at the (P')th place above each alphabet of the digraph in the Playfair matrix (going back to the very bottom if at the topmost respectively).

3. **If both the alphabets of the digraph are neither in the same column nor the same row:**
Then if (x, y) denotes the coordinate address of one of the alphabets of the digraph in the Playfair matrix, the coordinate address of the encrypted alphabet in the Playfair matrix, say (a, b), equals to

$$([x - \{P' + 1\}]modulo\ 5, [y - \{P' + 1\}]modulo\ 5).$$

Example: Keyword – "CAR"
Plaintext – "HEAL"
Secret key – 3

Using the process of encryption of the extended Playfair cipher, the plaintext is divided into digraphs: "HE, AL."

C	A	R	B	D
E	F	G	H	I
K	L	M	N	O
P	Q	S	T	U
V	W	X	Y	Z

Consider the digraph "HE." Clearly, both the letters are in the same row. So the letter "H" is encrypted to the letter "F" and the letter "E" is encrypted to the letter "H."

Consider the digraph "AL." Clearly, both the letters are in the same column. So the letter "A" is encrypted to the letter "Q" and the letter "L" is encrypted to the letter "A."

Combining both the encrypted digraphs, the ciphertext is "FHQA."

7.4.2 The diffie–hellman algorithm

The Diffie–Hellman algorithm is used to share the secret key between the users, which can further be used to exchange secret communication over a public network. The original implementation of the protocol uses four variables: one prime number, say F, a primitive root of F, say G, and two private values, respectively.

Both F and G are publicly shared. Users, say A and B, can pick their private key values, say if the private key of user A is *PrA* and the private key of user B is *PrB*. Then these private keys will be used to generate their public keys, say the public key of A is *PuA* and the public key of B is *PuB*. Then the following formula can be used to generate their public keys:

$$PuA = G^{PrA} \ modulo \ F$$

$$PuB = G^{PrB} \ modulo \ F.$$

Then by using the public key of A and the private key of B, user B can compute the value of the secret key shared by the algorithm. Similarly, by using the public key of B and the private key of A, user A can compute the value of the secret key shared by the algorithm. To compute the secret key, the following method can be used.

Secret key computed by A:

$$PuB^{PrA} \ modulo \ F.$$

Secret key computed by B:

$$PuA^{PrB} \ modulo \ F.$$

In the end, both the users will come up having the symmetric secret key to encrypt.

7.5 Proposed Work

The proposed work consists of the following four steps.

1. In the first step, use the Diffie–Hellman algorithm to exchange the secret key number between the sender and the receiver. Here, in this step, the chosen prime number P and primitive root G need to be publicly shared.
2. In the second step, a meaningless, publicly shared keyword will be encrypted to a private keyword using the secret key number from step 1. This private keyword will be used as the monarchy in the extended Playfair cipher in step 3.
3. In the third step, construct the Playfair matrix by using the private keyword (monarchy) from step 2, and follow the rules of the extended Playfair cipher (4.1) to construct the matrix.
4. In the last step, use the extended Playfair cipher (4.1) to encrypt or decrypt the message. Here, we will use the secret key number from step 1 as the private key P in the extended Playfair cipher.

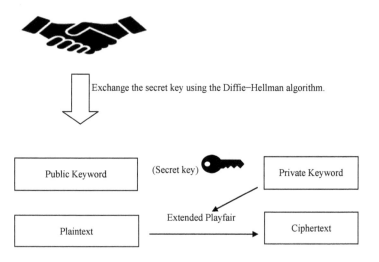

Note that the same secret key number from step 1 is being used in both step 2 and step 4, respectively.

7.6 Methodology

The methodology is divided into four parts.

Part 1:

Use the Diffie−Hellman algorithm (4.2) to find the secret key number.

Algorithm:

F − Prime number (publicly shared)

G − Primitive root of F (publicly shared)

PrA − Private key of user A

PrB − Private key of user B

PuA − Public key of user A

PuB − Public key of user B

SN − Secret key number

Then, by the original implementation of the Diffie−Hellman algorithm:

$$PuA = G^{PrA} \ modulo F$$

$$PuB = G^{PrB} \ modulo F.$$

After this, both the users can find the value of "SN" by using each other's public key by following:

$$SN = PuB^{PrA} \ modulo \ F$$

$$SN = PuA^{PrB} \ modulo \ F.$$

This SN will be symmetrically determined by both A and B.

Part 2:

Now, a meaningless, publicly shared "public keyword" needs to be encrypted to the private keyword (monarchy) for the Playfair matrix using the secret number "SN," by following the given rules.

First, determine that each of the 26 alphabets can be represented by their corresponding numerical address number. That is,

Let us denote corresponding number as CN.

Then, each alphabet of the public keyword can be encrypted by using the following operation:

$$[(CN \times SN) modulo \ 26] + 1.$$

This encryption will end up giving the private keyword (monarchy) for the extended Playfair cipher.

A-1	B-2	C-3	D-4	E-5	F-6	G-7	H-8	I-9	J-10	K-11	L-12	M-13
N-14	O-15	P-16	Q-17	R-18	S-19	T-20	U-21	V-22	W-23	X-24	Y-25	Z-26

Let us denote the private keyword for the extended Playfair cipher by "PKw."

Part 3:

Now, construct the extended Playfair square of the 5×5 matrix as we have discussed in Section 7.4.1. For constructing the matrix, make sure to use the private keyword "PKw" from the above part as the monarchy of the matrix.

** Make sure not to choose a very long public keyword because, in that way, one would end up with a very long private keyword that might fail to fit in the Playfair matrix of 25 entries.

Also, make sure not to repeat the alphabet in the public keyword. If for any reason alphabets get repeated in the private keyword, then use an alphabet once only, ignoring the further repetitions. **

Part 4:

Lastly, use the extended Playfair matrix constructed in Part 3 and the secret key number that is "SN" from Part 1 of Section 7.6.

In the algorithm of the extended Playfair cipher as discussed in Section 7.4.1, use "SN" in place of "P" (that is, use the secret key number "SN" as the private key "P" in the extended Playfair cipher) and use the extended Playfair matrix from the above part. Finally, the user can encrypt or decrypt the message using the encryption and decryption algorithms of the extended Playfair cipher (Section 7.4.1), respectively.

7.7 Illustration

~ Plaintext – "MILDER"
~ (Chosen prime number) F = 23
~ (Chosen primitive root of F) G = 5
~ (Private key chosen by user A) $PrA = 4$
~ (Private key chosen by user B) $PrB = 3$

Step 1:

Public key value of user A can be calculated as
~ $PuA = G^{PrA}$ modulo F = 5^4 modulo 23 = 625 modulo F = 4
~ $PuB = G^{PrB}$ modulo F = 5^3 modulo 23 = 125 modulo F = 10.

Secret key number SN computed by user A:
$SN = PuB^{PrA}$ modulo F = 10^4 modulo 23 = 10,000 modulo 23 = 18.

Secret key number SN computed by user B:

$SN = PuA^{PrB}$ modulo $F = 4^3$ modulo $23 = 64$ modulo $23 = 18$.

Clearly, the secret key number computed by both A and B are symmetric, which verifies that the steps are correctly done. Hence,

$\sim$ (Secret key number) SN = 18.

Step 2:

$\sim$ Public keyword – "HIBMT."

Then the private keyword (monarchy) "PKw" can be calculated as

(Corresponding address number) CN of the letter "H" = 8.

So, "H" is encrypted as

[(CN × SN) modulo 26] + 1 = [(8 × 18) modulo 26] + 1 = [144 modulo 26] + 1 = 14 + 1 = 15.

"15" is the CN of the letter "O." So, the letter "H" is encrypted to the letter "O," respectively.

In a similar way, CN of the letter "I" = 9, following the encryption process:

[(CN × SN) modulo 26] + 1 = [(9 × 18) modulo 26] + 1 = [162 modulo 26] + 1 = 6 + 1 = 7.

"7" is the CN of the letter "G." So, the letter "I" is encrypted to the letter "G," respectively.

CN of the letter "B" = 2, following the encryption process:

[(CN × SN) modulo 26] + 1 = [(2 × 18) modulo 26] + 1 = [36 modulo 26] + 1 = 10 + 1 = 11.

"11" is the CN of the letter "K." So, the letter "B" is encrypted to the letter "K," respectively.

CN of the letter "M" = 13, following the encryption process:

[(CN × SN) modulo 26] + 1 = [(13 × 18) modulo 26] + 1 = [234 modulo 26] + 1 = 0 + 1 = 1.

"1" is the CN of the letter "A." So, the letter "M" is encrypted to the letter "A," respectively.

CN of the letter "T" = 20, following the encryption process:

[(CN × SN) modulo 26] + 1 = [(20 × 18) modulo 26] + 1 = [360 modulo 26] + 1 = 22 + 1 = 23.

"23" is the CN of the letter "W." So, the letter "T" is encrypted to the letter "W," respectively.

After the complete process of encryption of the public keyword, we can conclude the private keyword, that is, PKw is "OGKAW," respectively.

$\sim$ (Private keyword) PKw – "OGKAW."

Step 3:

Let us now construct the extended Playfair matrix using the monarchy "PKw."

Step 4:

O	G	K	A	W
B	C	D	E	F
H	I	L	M	N
P	Q	R	S	T
U	V	X	Y	Z

Encryption of the plaintext.

~ (Private key for the extended Playfair cipher) P = SN = 18

~ (Calculated key from the private key) P' = (18 modulo 5) + 1 = 3 + 1 = 4

~ Plaintext − "MILDER"

Now, follow the encryption algorithm of the extended Playfair cipher (4.1):

(Pairs of two alphabets) Digraphs of the plaintext − "MI, LD, ER."

Consider the digraph "MI."

Clearly, both the alphabets of the digraph are in the same row of the Playfair matrix. Using the encryption algorithm of the extended Playfair cipher.

Encryption of the digraph "MI" can be done like the following.

The letter at the fourth right place of letter "M" is "L" and the letter at the fourth right place of the letter "I" is "H." So, the digraph "MI" is encrypted to "LH," respectively.

Consider the digraph "LD."

Clearly, both the alphabets of the digraph are in the same column of the Playfair matrix. Using the encryption algorithm of the extended Playfair cipher.

Encryption of the digraph "LD" can be done like the following.

The letter at the fourth place below the letter "L" is "D" and the letter at the fourth place below the letter "L" is "K." So, the digraph "LD" is encrypted to "DK," respectively.

Consider the digraph "ER."

Clearly, both "E" and "R" are not in the same column or the same row of the Playfair matrix.

The coordinate address of the letter "E" in the Playfair matrix is (2, 4) and the coordinate address of the letter "R" in the Playfair matrix is (4, 3).

The coordinate address of the corresponding encrypted letter of the letter "E" in the Playfair matrix is

$([(2 + P') \bmod 5] + 1, [(4 + P') \bmod 5] + 1)$

$= ([(2 + 4) \bmod 5] + 1, [(4 + 4) \bmod 5] + 1)$

$= ([6 \bmod 5] + 1, [8 \bmod 5] + 1) = (1 + 1, 3 + 1) = (2, 4).$

(2, 4) is the coordinate address of the letter "E." So the letter "E" is encrypted by the letter "E," respectively.

The coordinate address of the corresponding encrypted letter of the letter "R" in the Playfair matrix is

$([(4 + P') \text{ modulo } 5] + 1, [(3 + P') \text{ modulo } 5] + 1)$

$= ([(4 + 4) \text{ modulo } 5] + 1, [(3 + 4) \text{ modulo } 5] + 1)$

$= ([8 \text{ modulo } 5] + 1, [7 \text{ modulo } 5] + 1) = (3 + 1, 2 + 1) = (4, 3)$.

(4, 3) is the coordinate address of the letter "R." So, the letter "R" is encrypted by the letter "R," respectively.

Combining all the encrypted digraphs:

The ciphertext/encrypted text – "LHDKER," respectively.

Step 5:

Decryption of the ciphertext.

~ (Private key for the extended Playfair cipher) P = SN = 18

~ (Calculated key from the private key) $P' = (18 \text{ modulo } 5) + 1 = 3 + 1 = 4$

~ ciphertext – "LHDKER"

Now, follow the decryption algorithm of the extended Playfair cipher(4.1):

(Pairs of two alphabets) Digraphs of the ciphertext – "LH, DK, ER."

Consider the digraph "LH."

Clearly, both the alphabets of the digraph are in the same row of the Playfair matrix. Using the decryption algorithm of the extended Playfair cipher.

Decryption of the digraph "LH" can be done like:

The letter at the fourth left place of letter "L" is "M" and the letter at the fourth left place of the letter "H" is "I." So, the digraph "LH" is decrypted to "MI," respectively.

Consider the digraph "DK."

Clearly, both the alphabets of the digraph are in the same column of the Playfair matrix. Using the encryption algorithm of the extended Playfair cipher.

Decryption of the digraph "DK" can be done like:

The letter at the fourth place above the letter "D" is "L" and the letter at the fourth place above the letter "K" is "D." So, the digraph "DK" is decrypted to "LD," respectively.

Consider the digraph "ER."

Clearly, both "E" and "R" are not in the same column or the same row of the Playfair matrix.

The coordinate address of the letter "E" in the Playfair matrix is (2, 4) and the coordinate address of the letter "R" in the Playfair matrix is (4, 3).

The coordinate address of the corresponding decrypted letter of the letter "E" in the Playfair matrix is

([2 − (P' + 1)] modulo 5, [4 − (P' + 1)] modulo 5)

= ([2 − (4 + 1)] modulo 5, [4 − (4 + 1)] modulo 5)

= ([2 − 5] modulo 5, [4 − 5] modulo 5) = ([−3] modulo 5, [−1] modulo 5)

= (2, 4).

(2, 4) is the coordinate address of the letter "E." So the letter "E" is decrypted by the letter "E," respectively.

The coordinate address of the corresponding decrypted letter of the letter "R" in the Playfair matrix is

([4 − (P' + 1)] modulo 5, [3 − (P' + 1)] modulo 5)

= ([4 − (4 + 1)] modulo 5, [3 − (4 + 1)] modulo 5)

= ([4 − 5] modulo 5, [3 − 5] modulo 5) = ([−1] modulo 5, [−2] modulo 5)

= (4, 3).

(4, 3) is the coordinate address of the letter "R." So the letter "R" is decrypted by the letter "R," respectively.

Combining all the decrypted digraphs:

The plaintext/decrypted text − "MILDER," respectively.

7.8 Advantages of Securing Key in the Extended Playfair Cipher using the Diffie–Hellman Algorithm

1. Diffie–Hellman algorithm is known for its best attack proof nature and making use of this algorithm for key exchange in the Playfair cipher, and ends up making the cipher more secure.
2. Since there are infinitely many words that exist, estimating the keyword (monarchy) secured by the secret key is not feasible.
3. Since all the encryption and decryption processes in the Playfair cipher depend on the Playfair matrix, making the keyword (monarchy) of the matrix private enhances the security of the Playfair cipher.
4. Changing cipher techniques of the Playfair cipher by introducing the role of secret key for the movement of alphabets in the Playfair matrix makes the extended Playfair cipher itself more secure than the traditional Playfair cipher.

7.9 Conclusion

So far, the symmetric Playfair cipher working with the concept of using the same traditional way of doing encryption and decryption has been used. The technique was manual and very easy to break. Pointing out all the shortfalls of the Playfair cipher, we have introduced a completely new version, by calculating the private keyword for the Playfair matrix using the secret key secured by the Diffie–Hellman algorithm, and by changing the cipher technique in the Playfair cipher to a completely new extension by making use of the same secret key secured by the Diffie–Hellman algorithm. Overall, in this new extension, the Diffie–Hellman algorithm is being used to secure the secret key and completely new cipher techniques are introduced in the Playfair cipher.

Acknowledgement

None.

References

[1] S. Pednekar, R. D. Kulkarni and P. Mahanwar, "Soft Computing Techniques." Institute of Distance and Open Learning, University of Mumbai.

[2] J. S. Joseph, S. S. Samy and V. Haribaabu, "Advanced Encryption using Fuzzy logic and Secret sharing scheme." International Journal of Pure and Applied Mathematics, Volume 118 No. 22 2018, 1743-1748.

[3] K. GaneshKumar and D. Arivazhagan, "New Cryptography Algorithm with Fuzzy Logic for Effective Data Communication", Indian journal of Science and Technology, Vol 9(48), DOI: 10.17485/ijst/2016/v9i48/108970, December 2016.

[4] C. Paar and J. Pelzel, "Understanding Cryptography, A Textbook for Students and Practitioners."

[5] D. E. R. Denning, "Cryptography and Data Security", PURDUE UNIVERSITY.

[6] W. Stallings, "Cryptography and Network Security", Principles and Practice, Global Edition.

[7] S. Bhattacharyya, N. Chand and S. Chakraborty, "A Modified Encryption Technique using Playfair Cipher 10 by 9 Matrix with Six Iteration Steps." International Journal of Advanced Research in Computer Engineering & Technology (IJARCET), Volume 3, Issue 2, February 2014.

[8] V. Kalaichelvi, K. Manimozhi, P. Meenakshi, B. Rajakumar, V. Devi, "An Adaptive Playfair cipher Algorithm for Secure Communication using Radix 64 Conversion." International Journal of Pure and Applied Mathematics, Volume 117 No.20 2017, 325-330.

[9] G. Sharma; S. S. Kushwa, P. Goyal, "Implementation of Modified Play-fair CBC Algorithm," International Journal of Engineering Research & Technology (IJERT), ISSN: 2278-0181, Vol. 5 Issue 06, June-2016.

[10] R. Babu, S. U. Kumar, A. V. Babu, I. V. N. S Aditya, P. Komuraiah, "An Extension to Traditional Playfair Cryptographic Method." International Journal of Computer Applications (0975-8887), Volume 17-No.5, March 2011.

[11] Md. A. T. Shakil and Md. R. Islam, "An Efficient Modification to Playfair Cipher." Ulab Journal of Science and Engineering, vol.5, no.1, November 2014 (1ssn:2079-4398).

[12] P. Pal, G. S. Thejas, S. K. Ramani, S. S. Iyengar and N. R. Sunitha, "A Variation in the working of Playfair Cipher." Published 2019, 4th International Conference on Computational Systems and Information Technology for Sustainable Solution (CSITSS).

[13] S. Karthiga and T. Velmurugan, "Enhancing Security in Cloud Comput-ing using Playfair and Caesar Cipher in Substitution Techniques." International Journal of Innovative Technology and Exploring Engineering (IJITEE), ISSN: 2278-3075, Volume-9 Issue-4, February 2020.

[14] A. A. Alam, B. S. Khalid and C. M. Salam, "A Modified Version of Playfair Cipher Using 7×4 Matrix." International Journal of Computer Theory and Engineering, Vol. 5, No. 4, August 2013.

[15] S. S. Chauhan, H. Singh and R. N. Gurjar, "Secure Key Exchange using RSA in Extended Playfair Cipher Technique." International Journal of Computer Applications (0975-8887), Volume 104- No 15, October 2014.

8

Application of Multi-criteria Decision-making in Sustainable Resource Planning

Shristi Kharola[1], Aayushi Chachra[1], Mangey Ram[1,2], Akshay Kumar[3], and Nupur Goyal[1]

[1]Graphic Era Deemed to be University, India
[2]Institute of Advanced Manufacturing Technologies, Peter the Great St. Petersburg Polytechnic University, Russia
[3]Graphic Era Hill University, India

Abstract

In comparison to conventional energy systems, renewable energy systems provide various advantages and benefits. Due to a dearth of conventional energy resources, the world is moving toward renewable energy systems at a faster pace. Before considering renewable energy systems as a sustainable alternative to conventional energy systems, a few essential aspects need to be studied, which are beneficial to understand the value of renewable energy systems. The study aims to identify and evaluate these key aspects using existing research. However, due to ambiguity, these aspects may be equally or unequally influential. As a result, the article uses multi-criteria decision-making to evaluate these aspects according to their significance.

Keywords: fuzzy theory, fuzzy analytical hierarchy process, multi-criteria decision-making, Renewable energy systems.

8.1 Introduction

1.2 billion people, or roughly 17%, of the global population do not have access to electricity, with Africa accounting for 635 million and India accounting for 237 million. 2.7 billion people worldwide still rely on traditional energy sources such as solidified dung cakes, firewood, and so on to fulfill their energy requirements. Most critically, 95% of this population comes from rural locations where modern energy resources are unavailable [32]. Rising population and industrialization are causing rises in energy demand.

Renewable energy resources and non-renewable energy resources are the two basic categories of energy resources. Resources such as coal, oil, and natural gas are non-renewable resources, also called fossil fuels, whereas resources like wind, solar, geothermal, hydrogen and wave energy, hydraulics, and biomass are renewable energy resources. Energy is requisite for the economic development of any nation. For the sustainable development of any society, we need abundant energy sources [38]. Despite renewable resources being readily available, non-renewable energy systems (NRES) are being utilized massively all over the world for development in almost every field [39]. Unfortunately, non-renewable energy sources are finite and will exhaust after some time. Also, fossil fuels cause air pollution and, hence, increase the carbon footprint.

Renewable energy systems (RES) create less pollution and are much more sustainable than NRES. The resources are easily replenished and will never exhaust over time, making the systems more reliable. RES generates energy that produces no greenhouse gas emissions, which means zero carbon emissions. It diversifies the energy supply and, thus, reduces dependence on imported fuels [33]. There is no doubt that a complete transformation to RES will have more challenges; for example, they are highly expensive. The energy production is unpredictable as it might depend on the weather, surroundings, and environment, and they might still have a low carbon footprint (not as much as NRES), but it is still a viable option than moving on with NRES [42].

The purpose of this research is to rank the primary economic, technological, and environmental factors that make RES a viable alternative over NRES. Previously, several studies analyzed various factors in the context of various nations, as shown in the literature review section [31]. The goal of this research is to generalize and then examine the factors in order to assess the feasibility of a RES. A multi-criteria decision-making (MCDM) process is assigned as an evaluation structure for resolving these issues in energy planning. The study first characterizes and describes the individual factors under each major factor and then separately evaluates these factors to study the prioritization and ranking, based on the fuzzy analytical hierarchy process (FAHP) technique.

The remaining part of the study is organized as follows. Section 8.2 discusses previous research on factors discovered earlier. In Section 8.3, the MCDM technique and FAHP are presented. In Section 8.4, following the determination of the selection criteria, the proposed methodology is applied to the factors. Section 8.5 discusses the results obtained from the study followed by closing remarks in Section 8.6.

8.2 Literature Review

Table 8.1 Factors identified from the literature.

Major factors	Sub-factors		Literature
Economic factors	Operating and maintenance cost	(E_1)	[1]
	Return on investment	(E_2)	[3]
	Affordability	(E_3)	[4]
	Reduced energy bill	(E_4)	[5]
	Cost of FUEL	(E_5)	[6], [7]
Technological factors	Resource availability	(T_1)	[8]
	Output capacity	(T_2)	[9], [10]
	Reliability of the system	(T_3)	[11]
	Consistency in innovation	(T_4)	[12]
Environmental factors	Emission rates	(N_1)	[13]
	Land availability	(N_2)	[14]
	Resource renewability	(N_3)	[16]
	Sustainability	(N_4)	[17]

All sustainable energy development strategies assume that they will generate demand for climate-friendly technology that is now unavailable or inaccessible at optimal levels under current market conditions. In terms of energy resource sustainability, an RES with less than 100% life-cycle efficiency is still better than a fossil-based system. According to the literature, MCDM techniques in designing sustainable energy are frequently used and well-known. The study [34] shed light on how renewable energy resources should be prioritized for power generation in Pakistan. To evaluate and prioritize the development of five low-emission energy technologies in Poland, Ligus and Peternek [30] suggested a hybrid MCDM model based on FAHP and FTOPSIS. Criteria were established for this aim, based on the achievement of Poland's sustainable development policy goals. The findings of the study suggest that nuclear power should be replaced by renewable energy technology. Using the analytical hierarchy process (AHP) and FVIKOR methodologies, Anser *et al.* [15] classified numerous locations in Turkey based on economic, environmental, and social criteria. To establish the solar power project in rural locations of Turkey, several important criteria were merged through mathematical development. Zhang *et al.* [23] developed a multi-criteria assessment technique to assess the public and private sectors' effect on the use of renewable energy technologies in houses. The proposed approach was adapted by the authors to assess renewable technology in Lithuanian houses. Saleem and Ulfat [24] used the AHP technique to evaluate renewable energy technologies and prioritized those that would assist satisfied power demand in Pakistan's residential sector. Solar power has been identified as the greatest renewable energy source for Pakistan in order to address the problem of electricity shortages in the residential sector. Wind power came second, biomass came third, hydro came fourth, and ocean and geothermal energy came last.

Renewable energy sustainability, power quality issues, energy storage, and energy allocation are some of the major factors in the energy planning segment [21]. Choosing a place for a wind power facility has always been a difficult task. In an unsettled context, fuzzy logic is more likely to approach decision-making challenges [26]. The application of AHP and fuzzy logic for storage energy technology selection in the context of power quality has been investigated, using efficiency, load management, technical maturity, cost, and life-cycle as factors [27]. In [28], the functioning of energy storage is evaluated by prioritizing criteria and examining three alternative scenarios utilizing AHP and fuzzy logic. Meyar-Naimi and Vaez-Zadeh [22] offered a case study of Iran that employed AHP to examine the power generating

Table 8.2 Description of criteria.

Sub-factors	Description
Operating and maintenance cost	The early investment in RES is always higher than in NRES, but their operating and maintenance cost is comparatively less than non-renewable sources, making them more cost-effective and reliable.
Return on investment	The energy return on investment (EROI) is a ratio that represents the amount of usable energy gained for each unit of energy invested into the process of acquiring that energy. When EROI is low, more input energy is required to produce the same amount of output energy.
Affordability	Although RES creates less pollution and less greenhouse gas emissions, they have a very high initial cost due to which they are comparatively less affordable than NRES. Government schemes or technical innovations are required to make them affordable and, hence, reduce our dependence on NRES.
Reduced energy bill	Despite the high initial cost of RES, they are much more cost-effective in the long term. They help in reducing the energy bill compared to NRES. For example, wind energy is the least expensive resource among all the technologies available. Solar panels help in reducing energy costs dramatically and increase the overall energy supply. Non-renewable sources such as natural gas, coal, oil, etc., were of great importance for economic progress but are detrimental to the environment. RES creates less pollution, but even with a high initial cost, it can help in reducing energy bills.
Cost of fuel	Many nations are striving to increase renewable energy supply and decrease energy supply from non-renewable resources like oil and coal as they cannot be replenished and have a comparatively high price. Also, renewable sources are abundant in nature and provide at least 1000 times more energy than all fossil fuels combined. The cost of fuel is negligible compared to non-renewable energy sources.

Table 8.2   *(Continued).*

Resource availability	Renewable energy sources like solar, wind, water, etc., are abundant in nature, and unlike non-renewable sources, they do not deplete. Hence, the resource for RES is always available. Unlike fossil fuels, renewable sources can be replenished. The availability of various types of renewable resources like solar, wind, or water may vary from place to place, but it will be available abundantly.
Output capacity	The main issue faced with renewable sources is that their availability is fluctuating. It depends on the weather, thus affecting its output capacity. However, due to technological advancements, the output capacity of RES can be increased.
Reliability of the system	The reliability of RES can be increased using technological innovations such as using parallel combinations or redundancy of the systems, thus making them more reliable and effective for use even in isolated areas with low accessibility.
Consistency in innovation	Because of technological innovations in renewables, the cost is continuously declining and being promoted. It helps in dealing with climate change and other such problems. Innovations in RES have helped in making the systems more reliable and reducing their cost so that RES is widely used as compared to NRES.
Emission rates	The biggest problem with NRES is the high emission rates. On the other hand, RES has very low emission rates. Akella *et al.* [13] have shown that the emission rates have reduced exponentially after the installation of RES. Also, the emission reductions are measurable and long-term.
Land availability	Land availability depends on various constraints such as proximity to woodlands and agriculture is preferred for RES, whereas the lands closer to tourist places, mining areas, or camping sites are not preferable for RES.

Table 8.2 *(Continued).*

Resource renewability	Fossil fuels are less preferred compared to renewable sources like solar, water, wind, etc., because of their non-renewability even with carbon capture technologies. Also, issues such as climate change, sustainable development, and the protection of the environment increase the demand for renewable resources.
Sustainability	Energy is a prerequisite for sustainable development, and, thus, preference for renewable energy sources such as solar, geothermal, wind, etc., is high. For sustainability, a transition from non-renewable systems to renewable systems is taking place for the past few decades.

system from a sustainability viewpoint based on an upgraded policy-making framework.

Economic, technological, and environmental aspects are three of the major aspects that help to decide the value of RES over NRES [25]. Despite various factors involved in these major factors, this study has identified some major sub-factors that motivate the world to move from NRES to RES. These key factors have been identified from the existing literature and further evaluated using expert opinions followed by an MCDM technique, FAHP, to rank these factors in order to understand the relevance of each factor that leads to the world choosing RES as future sustainability goals. Table 8.1 briefs the major three factors and the sub-factors identified from the literature. Further, Table 8.2 describes each of these factors in detail.

8.3 Methodology

Given the numerous sustainability scenarios and elements, the MCDM model is most suited for such a revolutionary goal. To provide the best solution that addresses all environmental and local issues in a real-time application, the MCDM model must be applied to a variety of criteria, including numerous situations [2, 35]. AHP is basic, adaptable, and intuitive, and it can handle qualitative and quantitative criteria; yet, it gets more difficult when applied to a large number of criteria, as noted in [29].

Table 8.3   Scale of relative importance [37].

Linguistic variable	Scale	Triangular fuzzy number
Equally preferred	1	(1, 1, 1)
Moderately preferred	3	(2, 3, 4)
Strongly preferred	5	(4, 5, 6)
Very strongly preferred	7	(6, 7, 8)
Extremely preferred	9	(9, 9, 9)
Intermediate values	2, 4, 6, 8	(1, 2, 3); (3, 4, 5); (5, 6, 7); (7, 8, 9)

The AHP hierarchical structure is solved by combining local and global preference weights to determine the overall priority (Saaty 1980). The AHP is divided into three major phases [36]:

1. creating a hierarchy of objectives and criteria;
2. performing pairwise comparisons of criteria at each level;
3. prioritization to rate the options.

Because of limited information, inaccurate human judgments, and a fuzzy environment, the conventional AHP is insufficient for dealing with fuzziness and uncertainty in MCDM. As a result, the FAHP methodology may be seen as an improved analytical method that evolved from traditional AHP. The fuzzy AHP algorithm is composed of the nine phases listed below [43].

Step 1: Selection of criteria.

Step 2: Examining scale of relative importance.

Step 3: Calculating fuzzy scale using linguistics variables.

Step 4: Establishing pairwise comparison matrix.

Step 5: Transforming pairwise comparison matrix into judgment matrix.

Step 6: Determining fuzzy geometric mean r_i and fuzzy weights $w_{i.}$, calculated as

$$r_i = (a_{i1} \otimes a_{i2} \otimes \ldots \otimes a_{in})^{1/n}, \forall i = 1, 2\ldots, n \qquad (8.1)$$

w_i of the respective indexes are calculated by the following formula:

$$w_i = r_i \otimes (r_1 \oplus r_2 \oplus \ldots \oplus r_n)^{-1}. \qquad (8.2)$$

Step 7: Defuzzification of fuzzy weights.

Step 8: Normalizing the weights.

Step 9: Ranking criteria.

Each ratio reflecting the relative importance of a pair of elements is represented in a matrix from which appropriate weights may be selected.

We changed the above-mentioned procedure in such a manner that those decision-makers are requested to express their opinions in fuzzy numbers with triangle membership functions since these ratios are inherently fuzzy – they represent a decision-maker's perspective on the importance of a pair of elements. Table 8.3 describes the relative relevance scale used in this investigation.

8.4 Numerical Analysis

Tables 8.4–8.18 numerically solve the factors economic, technological, and environmental, respectively, based on the FAHP approach.

8.4.1 Economic factors

Table 8.4   Pairwise comparison matrix.

Economic factors	E_1	E_2	E_3	E_4	E_5
E_1	1	7	5	6	1/9
E_2	1/7	1	5	4	1/7
E_3	1/5	1/5	1	1/4	1/8
E_4	1/6	1/4	4	1	1/7
E_5	9	7	8	7	1

Table 8.5   Fuzzified pairwise comparison matrix.

Economic factors	E_1	E_2	E_3	E_4	E_5
E_1	(1, 1, 1)	(6, 7, 8)	(4, 5, 6)	(5, 6, 7)	(1/9, 1/9, 1/9)
E_2	(1/8, 1/7, 1/6)	(1, 1, 1)	(4, 5, 6)	(3, 4, 5)	(1/8, 1/7, 1/9)
E_3	(1/6, 1/5, 1/4)	(1/6, 1/5, 1/4)	(1, 1, 1)	(1/5, 1/4, 1/3)	(1/9, 1/8, 1/7)
E_4	(1/7, 1/6, 1/5)	(1/5, 1/4, 1/3)	(3, 4, 5)	(1, 1, 1)	(1/8, 1/7, 1/6)
E_5	(9, 9, 9)	(6, 7, 8)	(7, 8, 9)	(6, 7, 8)	(1, 1, 1)

Table 8.6   Calculation of fuzzy weights.

Economic factors	Geometric mean	Fuzzy relative weights
E_1	(1.6787, 1.8775, 2.0626)	(0.1779, 0.2187, 0.2673)
E_2	(0.7154, 0.8359, 0.9642)	(0.0758, 0.0975, 0.1249)
E_3	(0.2281, 0.2626, 0.3124)	(0.0242, 0.0306, 0.0405)
E_4	(0.4036, 0.4735, 0.5609)	(0.0427, 0.05523, 0.0727)
E_5	(4.6895, 5.1227, 5.5326)	(0.4972, 0.5975, 0.7171)

Table 8.7   De-fuzzification and normalization.

Economic factors	Relative weights	Normalized weights
E_1	0.2213	0.1826
E_2	0.2982	0.2460
E_3	0.0317	0.0261
E_4	0.0568	0.0468
E_5	0.6039	0.4983

Table 8.8   Economic factor ranking.

	Factor	Rank
E_5	Cost of fuel	1
E_2	Return on investment	2
E_1	Operating and maintenance cost	3
E_4	Reduced energy bill	4
E_3	Affordability	5

8.4.2 Technological factors

Table 8.9   Pairwise comparison matrix.

Technological factors	T_1	T_2	T_3	T_4
T_1	1	9	7	5
T_2	1/9	1	4	1/6
T_3	1/7	1/4	1	1/7
T_4	1/5	6	7	1

Table 8.10 Fuzzified pairwise comparison matrix.

Technological factors	T_1	T_2	T_3	T_4
T_1	(1, 1, 1)	(9, 9, 9)	(6, 7, 8)	(4, 5, 6)
T_2	(1/9, 1/9, 1/9)	(1, 1, 1)	(3, 4, 5)	(1/7, 1/6, 1/5)
T_3	(1/8, 1/7, 1/6)	(1/5, 1/4, 1/3)	(1, 1, 1)	(1/8, 1/7, 1/6)
T_4	(1/6, 1/5, 1/4)	(5, 6, 7)	(6, 7, 8)	(1, 1, 1)

Table 8.11 Calculation of fuzzy weights.

Technological factors	Geometric mean	Fuzzy relative weights
T_1	(3.8336, 4.2128, 4.5590)	(0.5194, 0.6284, 0.7557)
T_2	(0.4671, 0.5216, 0.5773)	(0.0633, 0.0778, 0.0957)
T_3	(0.2364, 0.2672, 0.3102)	(0.03202, 0.0398, 0.0514)
T_4	(1.4953, 1.7024, 1.9343)	(0.2026, 0.2539, 0.3206)

Table 8.12 De-fuzzification and normalization.

Technological factors	Relative weights	Normalized weights
T_1	0.6345	0.6259
T_2	0.0789	0.0778
T_3	0.0411	0.0405
T_4	0.2591	0.2556

Table 8.13 Technological factor ranking.

	Factor	Rank
T_1	Resource availability	1
T_4	Consistency in innovation	2
T_2	Output capacity	3
T_3	Reliability of the system	4

8.4.3 Environmental factors

Table 8.14 Pairwise comparison matrix.

Environmental factors	N_1	N_2	N_3	N_4
N_1	1	2	5	9
N_2	1/2	1	4	5
N_3	1/5	1/4	1	7
N_4	1/9	1/5	1/7	1

Table 8.15   Fuzzified pairwise comparison matrix.

Environmental factors	N_1	N_2	N_3	N_4
N_1	(1, 1, 1)	(1, 2, 3)	(4, 5, 6)	(9, 9, 9)
N_2	(1/3, 1/2, 1)	(1, 1, 1)	(3, 4, 5)	(4, 5, 6)
N_3	(1/6, 1/5, 1/4)	(1/5, 1/4, 1/3)	(1, 1, 1)	(6, 7, 8)
N_4	(1/9, 1/9, 1/9)	(1/6, 1/5, 1/4)	(1/8, 1/7, 1/6)	(1, 1, 1)

Table 8.16   Calculation of fuzzy weights.

Environmental factors	Geometric mean		Fuzzy relative weights	
N_1	(2.4494, 3.5676)	3.0801,	(0.23009, 0.7508)	0.52516,
N_2	(1.4142, 2.3403)	1.7783,	(0.13285, 0.3032, 0.4925)	
N_3	(0.6687, 0.9036)	0.7692,	(0.0628, 0.19016)	0.13115,
N_4	(0.21934, 3.8336)	0.2374,	(0.0206, 0.04047, 0.8068)	

Table 8.17   De-fuzzification and normalization.

Environmental factors	Relative weights	Normalized weights
N_1	0.50202	0.4085
N_2	0.30952	0.2518
N_3	0.12804	0.1042
N_4	0.28929	0.2354

Table 8.18   Environmental factor ranking.

	Factor	Rank
N_1	Emission rates	1
N_2	Land availability	2
N_4	Sustainability	3
N_3	Resource renewability	4

8.5 Results and Discussion

The rankings of factors obtained from the relative analysis are presented in Figures 8.1, 8.2 and 8.3.

According to the results, the factor "Cost of Fuel" is clarified as the first priority in the case of RES as an alternative to NRES under the economic factor ranking. It is followed by "Return on Investment" in the second place and "Operating & Maintenance Cost" in the third place of the ranking. If

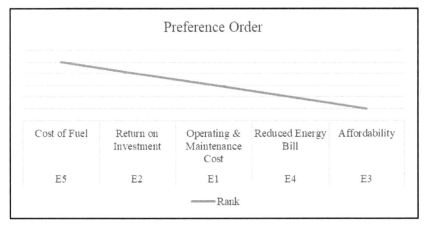

Figure 8.1 Economic factor ranking.

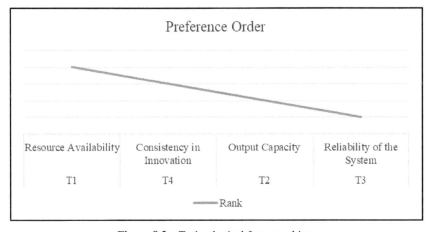

Figure 8.2 Technological factor ranking.

the world is to transition to low-carbon electricity, energy generated from these sources must be less expensive than power generated from fossil fuels. Because fossil fuel electricity was much cheaper than renewable electricity until recently, fossil fuel electricity has dominated the world power supply [18]. This has changed dramatically in the last 10 years. Power from new renewables is currently cheaper than power from new fossil fuels in most parts of the world. Renewable energy technologies follow learning curves, which means their prices reduce in proportion to the amount of cumulative

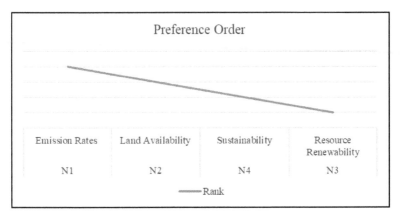

Figure 8.3   Environmental factor ranking.

installed capacity that doubles. However, because the price of energy gener-ated by fossil fuels does not follow a learning curve, we may anticipate the price gap between expensive fossil fuels and inexpensive renewables to widen more in the future [19]. Unlike conventional power plants, where fuel costs have traditionally dominated electricity generation costs (with the exception of nuclear power), the costs of renewable energy sources (RES) are primarily determined by their investment cost and local conditions, such as wind speed profiles or solar irradiation, as well as the cost of maintenance [7]. It should be emphasized, however, that all governments should begin investing in renewable energy sources far before 2030 in order to maximize this potential while minimizing negative consequences [20]. Under the technological factor ranking, the factor "Resource Availability" is classified as the first priority in the case of RES as an alternative to NRES, as shown in Figure 8.2. It is followed by "Consistency in Innovation" in the second place and "Output Capacity" in the third place of the ranking.

The availability of resources is crucial in resource management. System management must know which resources are accessible at any given time. Solar energy, hydropower, and geothermal energy should all be able to meet increasing demands for electricity in the years to come as they are abundant in nature.

Under the environmental factor ranking, the factor "Emission Rates" is classified as the first priority in the case of RES as an alternative to NRES. It is followed by "Land Availability" in the second place and "Sustainability" in the third place of the ranking.

"Reduced Energy Bill," "Affordability," "Reliability of the System," and "Resource Renewability" still lie at low-ranking levels as per the results obtained.

8.6 Conclusion

Due to the sheer rapid depletion of natural resources and unfavorable global environmental changes, it is critical to conserve natural resources and safeguard the environment. The need for energy is growing as the world's population grows. On the other side, the energy demands cannot be satisfied by the current energy sources. Renewable energy sources (RESs) are being looked at as an alternative since they are clean and ecologically friendly. The evaluation of variables influencing the choice of renewable energy alternatives over non-renewable energy alternatives is treated as an MCDM issue in this study, with the goal of determining the optimal factors influencing the choice of renewable energy alternatives. For this purpose, a fuzzy-based MCDM model for prioritizing variables impacting renewable energy sources has been proposed. The suggested model is built using the AHP approach with triangular fuzzy numbers.

The authors' MCDM approach was used to analyze the priority of economic, technical, and environmental aspects impacting the development of RES over NRES. After defining the weights of certain elements, the priority factors might be indicated in the form of a ranking. As RES options, 3 primary criteria (in this case, factors) and 13 sub-criteria (sub-factors) are considered. As a result, the proposed fuzzy-based MCDM model identifies the factors "Cost of Fuel," "Resource Availability," and "Emission Rates" as the best three factors under the respective major factors that enable the use of renewable energy alternatives, with relative weight values of 0.4983, 0.6259, and 0.4085, respectively. The findings and statistics have been examined to better implicate the findings. Different integrated fuzzy MCDM approaches can be employed to tackle this problem in the future, and the results can be compared to the results of this article.

Acknowledgement

The authors are grateful to reviewers, editors, publishers, and also to "Graphic Era Deemed to be University, Dehradun, Uttarakhand, India," for their support. This chapter was written in good faith using information available at the time of publication.

References

[1] M. K. Deshmukh and S. S. Deshmukh, 'Modeling of hybrid renewable energy systems.' Renewable and Sustainable Energy Reviews, vol. 12, no. 1, pp. 235-249, 2008.

[2] T. Ertay, C. Kahraman and İ. Kaya, 'Evaluation of renewable energy alternatives using MACBETH and fuzzy AHP multicriteria methods: the case of Turkey.' Technological and Economic Development of Economy, vol. 19, no. 1, pp. 38-62, 2013.

[3] L. C. King and J. C. Van Den Bergh, 'Implications of net energy-return-on-investment for a low-carbon energy transition.' Nature Energy, vol. 3, no. 4, pp. 334-340, 2018.

[4] X. L. Yuan and J. Zuo, 'Pricing and affordability of renewable energy in China–A case study of Shandong Province.' Renewable Energy, vol. 36, no. 3, pp.1111-1117,2011.

[5] A. V. Herzog, T. E. Lipman, J. L. Edwards and D. M. Kammen, 'Renewable energy: a viable choice.' Environment: Science and Policy for Sustainable Development, vol. 43, no. 10, pp. 8-20, 2001.

[6] H. Lund and B. V. Mathiesen, 'Energy system analysis of 100% renewable energy systems—The case of Denmark in years 2030 and 2050.' Energy, vol. 34, no. 5, pp. 524-531, 2009.

[7] S. R. Bull, 'Renewable energy today and tomorrow.' Proceedings of the IEEE, vol. 89, no. 8, pp. 1216-1226, 2001.

[8] B. Johansson, 'Security aspects of future renewable energy systems–A short overview.' Energy, vol. 61, pp. 598-605, 2013.

[9] C. Wang, B. Yu, J. Xiao and L. Guo, 'Sizing of energy storage systems for output smoothing of renewable energy systems.' Proceedings of the CSEE, vol. 32, no. 16, pp. 1-8, 2012.

[10] K. Mitchell, M. Nagrial and J. Rizk, 'Simulation and optimisation of renewable energy systems.' International Journal of Electrical Power & Energy Systems, vol. 27, no. 3, pp. 177-188, 2005.

[11] L. A. D. S. Ribeiro, O. R. Saavedra, S. L. Lima, J. G. de Matos and G. Bonan, 'Making isolated renewable energy systems more reliable.' Renewable Energy, vol. 45, pp. 221-231, 2012.

[12] Z. X. He, S. C. Xu, Q. B. Li and B. Zhao, 'Factors that influence renewable energy technological innovation in China: a dynamic panel approach.' Sustainability, vol. 10, no. 1, pp. 124, 2018.

[13] A. K. Akella, R. P. Saini and M. P. Sharma, 'Social, economical and environmental impacts of renewable energy systems.' Renewable Energy, vol. 34, no. 2, pp. 390-396, 2009.

[14] D. S. Ryberg, M. Robinius and D. Stolten 'Evaluating land eligibility constraints of renewable energy sources in Europe.' Energies, vol. 11, no. 5, pp. 1246, 2018.

[15] M. K. Anser, M. Mohsin, Q. Abbas and I. S. Chaudhry, 'Assessing the integration of solar power projects: SWOT-based AHP–F-TOPSIS case study of Turkey.' Environmental Science and Pollution Research, vol. 27, no. 25, pp. 31737-31749, 2020.

[16] I. Dincer and C. Acar 'Smart energy systems for a sustainable future.' Applied Energy, vol. 94, pp. 225-235, 2017.

[17] P. A. Østergaard, N. Duic, Y. Noorollahi, H. Mikulcic, S. Kalogirou, 'Sustainable development using renewable energy technology.' Renewable Energy, vol. 146, pp. 2430-2437, 2020.

[18] L. R. Brown, 'The Great Transition: Shifting from Fossil Fuels to Solar and Wind Energy.' The Humanist, vol. 76, no. 1, 2016.

[19] . Christiansson, 'Diffusion and learning curves of renewable energy technologies, 1995.

[20] M. Ram, M. Child, A. Aghahosseini, D. Bogdanov, A. Lohrmann, and C. Breyer, 'A comparative analysis of electricity generation costs from renewable, fossil fuel and nuclear sources in G20 countries for the period 2015-2030.' Journal of Cleaner Production, vol. 199, pp. 687-704, 2018.

[21] J. Siskosa and P. Hubert, 'Multi-criteria analysis of the impacts of energy alternatives: a survey and a new comparative approach.' European Journal of Operational Research, vol. 13, no. 3, pp. 278-299, 1983.

[22] H. Meyar-Naimi and S. Vaez-Zadeh, 'Sustainability assessment of Iran power generation system using DSR HNS framework.' In 2012 Second Iranian Conference on Renewable Energy and Distributed Generation (pp. 98-103), IEEE, March 2012.

[23] C. Zhang, Q. Wang, S. Zeng, T. Baležentis, D. Štreimikienė, I. Ališauskaitė-Šeškienė and X. Chen, 'Probabilistic multi-criteria assessment of renewable micro-generation technologies in households.' Journal of Cleaner Production, vol. 212, pp. 582-592, 2019.

[24] L. Saleem and I. Ulfat, 'A multi criteria approach to rank renewable energy technologies for domestic sector electricity demand of Pakistan.' Mehran University Research Journal of Engineering & Technology, vol. 38, no. 2, pp. 443-452, 2019.

[25] A. I. Chatzimouratidis and P. A. Pilavachi, 'Technological, economic and sustainability evaluation of power plants using the Analytic Hierarchy Process.' Energy Policy, vol. 37, no. 3, pp. 778-787, 2009.

[26] W. Ying-Yu and Y. De-Jian, 'Extended VIKOR for multi-criteria decision making problems under intuitionistic environment.' In 2011 International Conference on Management Science & Engineering 18th Annual Conference Proceedings (pp. 118-122), IEEE. September 2011.

[27] A. Barin, L. N. Canha, A. da Rosa Abaide and K. F. Magnago, 'Selection of storage energy technologies in a power quality scenario—the AHP and the fuzzy logic.' In 2009 35th Annual Conference of IEEE Industrial Electronics (pp. 3615-3620), IEEE. 2009a, November.

[28] A. Barin, L. N. Canha, A. da Rosa Abaide, K. F. Magnago and R. Q. Machado, 'Storage energy management with power quality concerns the analytic hierarchy process and the fuzzy logic.' In 2009 Brazilian Power Electronics Conference (pp. 225-231), IEEE.2009b, November.

[29] R. Ramanathan and L. S. Ganesh, 'Energy resource allocation incorporating qualitative and quantitative criteria: An integrated model using goal programming and AHP.' Socio-Economic Planning Sciences, vol. 29, no. 3, pp. 197-218, 1995.

[30] M. Ligus and P. Peternek, 'Determination of most suitable low-emission energy technologies development in Poland using integrated fuzzy AHP-TOPSIS method.' Energy Procedia, vol. 153, pp. 101-106, 2018.

[31] B. K. Sovacool, 'Design principles for renewable energy programs in developing countries.' Energy & Environmental Science, vol. 5, no. 11, pp. 9157-9162, 2012.

[32] I. E. Agency, 'World Energy Outlook 2015– Electricity Access Database,' 2015.

[33] A. Mardani, A. Jusoh, E. K. Zavadskas, F. Cavallaro, and Z. Khalifah, 'Sustainable and renewable energy: An overview of the application of multiple criteria decision making techniques and approaches.' Sustainability, vol. 7, no. 10, pp. 13947-13984, 2015.

[34] Y. A. Solangi, Q. Tan, N. H. Mirjat, G. D. Valasai, M. W. A. Khan and M. Ikram, 'An integrated Delphi-AHP and fuzzy TOPSIS approach toward ranking and selection of renewable energy resources in Pakistan.' Processes, vol. 7, no. 2, pp. 118, 2019.

[35] A. Kumar, B. Sah, A. R. Singh, Y. Deng, X. He, P. Kumar and R. C. Bansal, 'A review of multi criteria decision making (MCDM) towards sustainable renewable energy development.' Renewable and Sustainable Energy Reviews, vol. 69, pp. 596-609, 2017.

[36] S. Tesfamariam and R. Sadiq, 'Risk-based environmental decision-making using fuzzy analytic hierarchy process (F-AHP).' Stochastic Environmental Research and Risk Assessment, vo. 21, no. 1, pp. 35-50, 2006.

[37] F. A. Alzahrani, 'Fuzzy Based Decision-Making Approach for Estimating Usable-Security of Healthcare Web Applications.' CMC-Computers Materials & Continua, vol. 66, no. 3, pp. 2599-2625, 2021.

[38] A. M. Omer, 'Energy, environment and sustainable development.' Renewable and Sustainable Energy Reviews, vol. 12, no. 9, pp. 2265-2300, 2008.

[39] M. Jefferson, 'Accelerating the transition to sustainable energy systems.' Energy Policy, vol. 36, no. 11, pp. 4116-4125, 2008.

[40] S. M. Shaahid and M. A. Elhadidy, 'Opportunities for utilization of stand-alone hybrid (photovoltaic+ diesel+ battery) power systems in hot climates.' Renewable Energy, vol. 28, no. 11, pp. 1741-1753, 2003.

[41] J. J. Buckley, 'Fuzzy Hierarchical Analysis.' Fuzzy Sets and Systems, vol. 17, no. 3, pp. 233-247, 1985.

9

Fuzzy Logic based Decision Systems in the Healthcare Sector

V. Mittal, P. Malik, K. C. Purohit, and M. Ram

Graphic Era (Deemed to be) University, India

Abstract

Fuzzy logic is a powerful tool for real-world problems in which data is approximate but not fixed. Fuzzy models are capable of recognizing, representing, manipulating, understanding, and exploiting ambiguous and uncertain facts and information. The healthcare sector plays a very important role in our everyday lives, consisting of a high level of ambiguity in terms of our health and sickness. Disease diagnosis is the result of clinical reasoning based on the investigation of obtained data in medicine. Handling uncertainty and inaccuracies in the diagnosis process of disease is critical since it can aid healthcare workers in disease control. The fuzzy logic technique was considered the most suitable method to represent the uncertainty in diagnosis due to the absence of extensive knowledge, as well as a shortage of time and the fuzzy nature of disease diagnosis. This chapter includes various approaches proposed by the authors in the diagnosis process of various diseases like kidney disease, lung cancer, breast cancer, prostate cancer, etc.

Keywords: Fuzzy logic, fuzzification, de-fuzzification, disease diagnosis, healthcare, kidney disease, lung disease, heart disease.

9.1 Introduction

In circumstances where the data given is approximation-based rather than fixed and specific, fuzzy logic (FL) has emerged as a powerful and effective tool for tackling complicated real-world problems [1]. It should be considered a strong tool for real-time decision-making systems and other artificial intelligence applications. In real life, fuzzy systems are employed in a variety of applications [2]. Fuzzy systems can be found in elevators, razors, subways, medical gadgets, bond grading systems, and risk analysis systems for bank credit applications. In the medical field, there are various levels of uncertainty and imprecision; therefore, there is a wide scope for fuzzy systems [3, 4].

Although FL was initially presented in 1965 to handle ambiguity, there are assertions that it was already there by the turn of the century, based on statements like "I have figured out the logic of vagueness with something like completeness" by philosopher Charles Sanders Peirce in 1905 [5, 6]. "Uncertainty," "vagueness," and "fuzziness" are common phrases connected with FL, and they all have the same quality of not being clear or well-defined enough to be handled by hard computing methods [7]. As per previous literature, FL can be defined in many ways. It can be characterized as the attempt of mimicking the reasoning model followed by a human. These reasoning models make use of linguistic variables and concepts. Instead of crisp numbers and equations, the computer may process language variables and their degrees of membership [8, 9].

Fuzzy models are capable of recognizing, representing, manipulating, understanding, and exploiting ambiguous and uncertain facts and information. It is correct as they all are the examples of uncertain english. FL is a technique for coping with non-statistical causes of imprecision and uncertainty. It is a cutting-edge technique that combines engineering expertise with traditional system design [10]. FL can be used to build anything that was previously produced using traditional methods. In some circumstances, though, traditional solutions are easier, faster, and more efficient. The key to making FL work is to combine it with other strategies in a clever way. FL wants to create multivalued logic that can mimic human reactions to continual choices [11, 12].

9.2 Fuzzy Sets

In traditional crisp set theory, an element is either a member of a set or not. The fact that fuzzy set theory allows for intermediate grades of membership distinguishes it from standard crisp set theory. Let us assume $F = \{x\}$ as a group of objects. A fuzzy set A in F can be defined as a collection of ordered pairs and can be written as shown in the following equation:

$$A = \{(x, \mu_A(x)), x \in F\}, \tag{9.1}$$

where μ_A is the degree of membership of x in A [9]. The concept of fuzzy sets is elaborated with the following example. For the set of tall men, every person can construct a fuzzy set with unlike members. A man with a height of 100-150 cm is considered tall, while a man with a height of 180 cm is not tall. This demonstrates the subjective, case-dependent, and context-dependent nature of fuzzy sets.

In most instances related to medical science, it is tough to obtain specific definitions for the concepts of the medical field and relations between these ideas; so fuzzy sets provide an efficient method to represent such cases [13, 14].

9.2.1 Membership function

he fuzzy set is fully defined by the membership function. The degree of resemblance between an element and a fuzzy set is measured using a membership function. The membership function can be chosen randomly by the user based on his experience. Alternatively, it might be created using machine learning techniques like genetic algorithms or neural network techniques.

A value is allocated for each element x of input space by the membership function and is denoted by the symbol $\mu(x)$. The membership function converts the value of the input into its membership degree. The values returned by the membership function must be in the [0, 1] range. The fuzzy set F, where $F = \{(2, 0.4), (6, 0.1), (8, 1)\}$, is represented using the fuzzy notation $F = \{0.4/2, 0.1/6, 1/8\}$. In a fuzzy set, the membership degree can never be zero. The membership degree of the set F at 2 is represented in the following equation:

$$\mu F(2) = 0.4. \tag{9.2}$$

The meaning of the membership is determined by the assessment of the fuzziness membership function. The most commonly used shapes for representing membership functions are triangular, trapezoidal, linear, and Gaussian. However, the kind of membership function to be used is still a research issue. In a study by McNeill and Thro [15] and Duraisamy *et al.* [16], the triangle membership function is the most functional and often utilized function because of its simplicity.

9.2.2 Components of a fuzzy logic system

The FL systems are rule-based systems that make use of fuzzy theories to handle uncertainties [17]. Figure 9.1 shows all four components of an FL system.

a) Fuzzifier: This component of the FL is responsible for transforming the raw crisp input values into fuzzy values. The process of fuzzification is carried out by the fuzzifier.

The technique of transforming an item from crisp into a fuzzy set, and assigning a value of membership function for linguistic variables, is known as fuzzification. Subsequently, the handling of data in an FLC is built on the theory of fuzzy sets; early fuzzification is required and desired [18].

Figure 9.1 shows the fuzzification process for the temperature at $t = 12$ into the linguistic variables cold, normal, and warm. To find the membership value μ (12), the vertical line from X-axis is drawn at $t = 12$ and its intersection with the horizontal axis through Y-axis is calculated. The value found at Y-axis will be the membership value and is shown in eqn (9.3). According to Figure 9.1,

$$\mu(12) = 0.40. \tag{9.3}$$

b) Inference engine and knowledge base: The resultant fuzzy sets are analyzed in the inference engine as per the rules of the knowledge base after

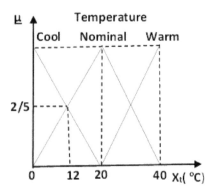

Figure 9.1   Fuzzification for temperature at $t = 12$.

fuzzification. The inference engine actually works like the processing unit of fuzzy systems.

The knowledge base is a very important component of any FL system. It determines the efficacy of a fuzzy system. It is based on the fuzzy set theory and provides the facts of the rules and linguistic variables in FL systems, allowing approximate reasoning [18]. An FL system's knowledge base is composed of a database and a rule base. The rule base consists of if−then rules, whereas the database specifies the fuzzy set membership functions that are utilized in fuzzy rules.

To construct a knowledge base, both the domain experts and the self-learning algorithm can be used. The domain experts define the if−then rules of the proposed fuzzy system. The self-learning fuzzy systems are also known as neuro-fuzzy systems [19]. For such systems, a portion of data is used for training purposes and to form the rule base, while the remaining data is intended to be solved by the system.

c) **Defuzzifier:** In most cases, the inference engine's output is likewise a fuzzy set, which is useless in real life. As a result, it must be translated into a practical and intelligible value that can be applied in the actual world. It is done by using the process called defuzzification. It is also known as the transformation of a fuzzy value into a crisp value. In many engineering applications, the defuzzification of the result is critical. To describe the process of defuzzification term, "Rounding it off" can be used [18]. Basically, it is the process of converting a fuzzy amount to an exact figure, while fuzzification is the process of converting an exact figure to a fuzzy number. Few methods that can be used to defuzzify the results are mentioned in the following.

1. **Max-membership method:** This method, commonly referred to as the height method, is confined to peak output functions. The algebraic expression shown in eqn (9.4) specifies this method:

$$\mu\left(f^*\right) \geq \mu(f) \forall f \in F \tag{9.4}$$

where f^* is the defuzzified value. It is shown in Figure 9.2.

2. **Centroid method:** The center of mass, center of the area, and center of gravity are the terms used to describe this strategy. It is the most used defuzzification technique. The defuzzified output f^* can be evaluated using the following equation:

$$f^* = \frac{\int \mu(f).f df}{\int \mu(f) df}. \tag{9.5}$$

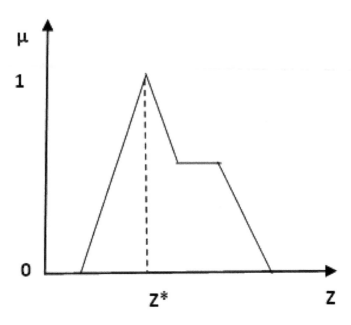

Figure 9.2   Max-membership method

The defuzzified output $f*$ using centroid method is shown in Figure 9.3.

3. **Weighted average method:** Only symmetrical output membership functions are valid for this approach. The maximum membership value is used to weigh each membership function. In this situation, the output is given by the following equation:

$$f^* = \sum \mu\left(f'\right) \cdot f'/\mu\left(f'\right), \qquad (9.6)$$

where f' denotes the maximum value in membership function and is shown in Figure 9.4.

4. **Mean-max membership:** The center of the maxima method is another name for this procedure. This is similar to the max-membership approach; other than that, the maximum membership positions might be non-unique. The output in the mean-max membership method is evaluated by the following equation:

$$f^* = \sum f'/n. \qquad (9.7)$$

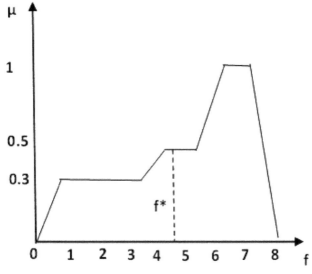

Figure 9.3   Centroid method.

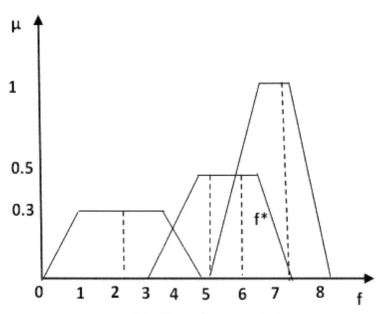

Figure 9.4   Weighted average method.

Here, f' is the maximum value of the used membership function and is shown in Figure 9.5.

5. **Center of sums:** Instead of combining the separate fuzzy subsets, this technique uses their algebraic total. The calculations are quick, but the intersecting areas are added twice, which is a downside as shown in Figure 9.6. The output is evaluated using the following equation:

$$f^* = \frac{\int f^* \mu(f).f df}{\int \mu(f) df}.$$ (9.8)

6. **Center of largest area:** When the output of at least two convex fuzzy subsets is not overlapping, this method can be used. In this situation, the output is skewed toward one side of a membership function. When the resulting fuzzy set includes at least two convex areas, the defuzzified value f^* is calculated using the center of gravity of the convex fuzzy subregion with the biggest area. The result of this method is given in the following equation:

$$f^* = \frac{\int \mu c(f).f df}{\int \mu c(f) df}.$$ (9.9)

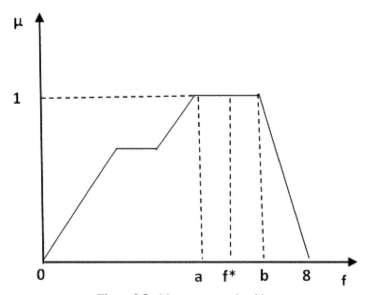

Figure 9.5   Mean-max membership.

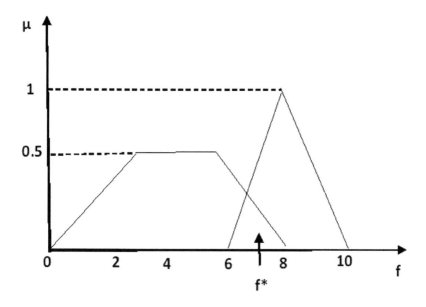

Figure 9.6 Centre of sum.

9.2.3 Operations on fuzzy sets

Let us assume two fuzzy sets U as the universe of information, x as an element of U, and $\mu(x)$ as the degree of membership of x. The following relation denotes the basic operations on the fuzzy sets:

1. **a) Union:**

$$\mu_{\tilde{A}\cup\tilde{B}}(x) = \mu_{\tilde{A}}\upsilon\mu_{\tilde{B}}\forall x \in U, \tag{9.10}$$

where denotes the max function. Union operation can also be known as OR operation as given in eqn (9.10) and is shown in Figure 9.7.

2. **b) Intersection:**

$$\mu_{\cup\tilde{B}}(x) = \mu\wedge\mu_{\tilde{B}}\forall x \in U, \tag{9.11}$$

where denotes the min function. The intersection also represents the AND operation as given in eqn (9.11) and is shown in Figure 9.8.

3. **c) Compliment:**

$$\mu = 1 - \mu(x)\forall x. \tag{9.12}$$

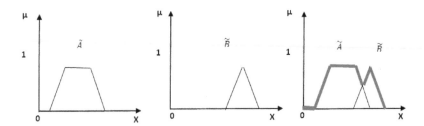

Figure 9.7 Union of two fuzzy sets.

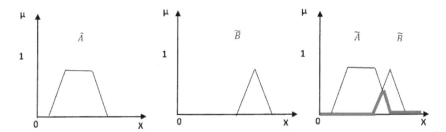

Figure 9.8 Intersection of two fuzzy sets.

It is similar to the NOT operation expressed in eqn (9.12) and is shown in Figure 9.9.

9.2.4 Fuzzy set properties

Some important properties of fuzzy sets are described in the following.

1. **Commutativity:** Consider two fuzzy sets $\tilde{A}$ and $\tilde{B}$. Then the commutative property shown in eqn (9.13) states that

$$\tilde{A} \cup \tilde{B} = \tilde{B} \cup \tilde{A}$$
$$\tilde{A} \cap \tilde{B} = \tilde{B} \cap \tilde{A}. \tag{9.13}$$

2. **Associativity:** Consider the fuzzy sets $\tilde{A}$, $\tilde{B}$ and $\tilde{C}$. Then the associative property shown in eqn (9.14) states that

$$(\tilde{A} \cup \tilde{B}) = \tilde{A} \cup (\tilde{B} \cup \tilde{C})$$
$$(\tilde{A} \cup \tilde{B}) = \tilde{A} \cap (\tilde{B} \cap \tilde{C}) \tag{9.14}$$

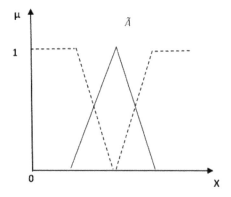

Figure 9.9 Compliment of a fuzzy set.

3. **Distributive:** Consider the fuzzy sets $\tilde{A}$, $\tilde{B}$ and $\tilde{C}$. Then the distributive property shown in eqn (9.15) states that

$$\tilde{A} \cup (\tilde{B} \cap \tilde{C}) = (\tilde{A} \cup \tilde{B}) \cap (\tilde{A} \cup \tilde{C})$$
$$\tilde{A} \cap (\tilde{B} \cup \tilde{C}) = (\tilde{A} \cap \tilde{B}) \cup (\tilde{A} \cap \tilde{C}). \tag{9.15}$$

4. **Idempotence:** Consider the fuzzy set $\tilde{A}$. Then the idempotent property given in eqn (9.16) states that

$$\tilde{A} \cup \tilde{A} = \tilde{A}$$
$$\tilde{A} \cap \tilde{A} = \tilde{A}. \tag{9.16}$$

5. **Identity:** Consider the fuzzy set $\tilde{A}$ and the universal set U. Then identity property is expressed through the following equation:

$$\tilde{A} \cup \emptyset = \tilde{A}$$
$$\tilde{A} \cap U = \tilde{A}$$
$$\tilde{A} \cap \emptyset = \varnothing$$
$$\tilde{A} \cup U = U. \tag{9.17}$$

6. **Transitivity:** Consider the fuzzy sets $\tilde{A}$, $\tilde{B}$ and $\tilde{C}$. Then the transitive property is shown in the following equation:

$$\text{if } \tilde{A} \subseteq \tilde{B} \subseteq \tilde{C}, \text{ then } \tilde{A} \subseteq \tilde{C}. \tag{9.18}$$

7. **Involution property:** Consider the fuzzy set $\tilde{A}$. Then the involution property given in eqn (9.19) states that

$$\overline{\overline{\tilde{A}}} = \tilde{A}. \tag{9.19}$$

8. **De Morgan's law:** Consider the fuzzy sets $\tilde{A}$ and $\tilde{B}$. The De Morgan's law in eqn (9.20) states that

$$\overline{\tilde{A} \cup \tilde{B}} = \overline{\tilde{A}} \cap \overline{\tilde{B}}$$
$$\overline{\tilde{A} \cap \tilde{B}} = \overline{\tilde{A}} \cup \overline{\tilde{B}}. \tag{9.20}$$

9.3 Fuzziness in Medical Field

The term "fuzzy" denotes a state of being perplexed and unclear; however, fuzziness should not be confused with vagueness. Because all reality in the world is tinged with ambiguity and uncertainty, no one can make an informed judgment based on insufficient information. To put it another way, the healthcare sector plays a very important role in our everyday lives, consisting of a high level of ambiguity in terms of our health and sickness. Disease diagnosis is the result of clinical reasoning based on the investigation of obtained data in medicine [20]. It entails several steps. To put it another way, every phase of the diagnosis might be a cause of doubt. The first phase is to collect data from the medical history of the patient, laboratory data, and data from other medical diagnostic tools. Despite the introduction of new diagnostic procedures and technology, the inherent ambiguity has not been reduced, and the complexity and uncertainty of final diagnosis have increased due to a large amount of data. As a result, physicians are frequently forced to make decisions by analyzing the gathered data once feasible in the iterative hypothetic deductive process.

Inadequate time for acquiring data may be the most significant impediment to making informed decisions. Errors while diagnosing, on the other hand, are unavoidable. However, these errors might occur as a result of a lack of expertise, a misdiagnosis, a lack of data collection, and clinical cognitive processes that underpin clinician diagnostic thinking.

It is critical to deal with uncertainty and inaccuracies in the diagnosis process of disease since it can aid healthcare workers in disease control. The FL technique was developed as a promising technique to represent the diagnosis ambiguity due to the absence of extensive knowledge, as well as a shortage of time and the fuzzy nature of disease diagnosis [6, 21, 22]. In this case, the sickness can be stated with a degree of membership to diagnosis using FL. In contrast to classical logic, where reality can only be characterized in terms of either true or false values, in FL, the ailment can be stated as a result of the diagnosis using intermediate truth values between right and wrong [23, 24]. The most frequent FL methods employed in the area of disease diagnosis will be briefly discussed in the following section.

9.3.1 Fuzzy in clinical decision support system (CDSS)

The majority of medical diagnoses and treatments are accompanied by a decision-making process. It is difficult to make timely and precise decisions based on the clinical expertise and medical history of the patient at the time of diagnosis, treatment, and management of diseases. CDSSs are computerized developed systems recognized in the healthcare sector to recommend the best option to aid clinicians in making decisions [25].

In the context of the decision support system, a more accurate mathematical model can be built by using FL to model physician diagnosis judgments. Knowledge representation is one of the most significant aspects of the decision support system, and FL can help solve the problem of knowledge representation with uncertainty and inaccuracy [26]. CDSSs are becoming increasingly important in enhancing the quality of therapeutic and diagnostic efficiency in healthcare. As a result of their high accuracy and low complexity, CDSSs are regarded as simple-to-use applications.

9.3.2 Fuzzy inference system (FIS) in disease diagnosis

The FIS is a system that is based on rules. It employs FL to represent various types of information about an issue and to describe the interactions and relationships that emerge between its variables. Through FIS, basically, the if−then rules are implemented. The FIS is a language framework for modeling human thought processes [27].

FIS for disease detection was designed to diagnose diseases more accurately and quickly. To build relationships between input and output data, FISs use a variety of fuzzy methodologies such as fuzzy membership sets and fuzzy-rule-based reasoning. Generally, numerous systems have been built using FIS approaches for the diagnosis of various diseases to date. The cornerstone of such fuzzy systems is that they are simple to set up, maintain, and operate. They are also inexpensive [28, 29].

9.3.3 Fuzzy classification and clustering

Due to rapid technological advancements, it is becoming increasingly difficult to analyze correct diagnoses without ambiguity using classic categorization methods. Because medical illnesses are often ambiguous, fuzzy techniques are more helpful than precise ones. For example, a symptom is a physical or laboratory finding that suggests the existence of a disease and hence serves as diagnostic assistance. A syndrome is a collection of symptoms that describe the existence and characteristics of an illness. One of the most important jobs in medical diagnosis is to classify the symptoms into a single syndrome. Clustering analysis is well-known in medicine as an effective and efficient method in this regard. A commonly used clustering algorithm in medical diagnosis is C-means clustering [30].

In various medical domains, fuzzy categorization has been utilized to extract a prediction model. A classification technique can aid clinicians in their decision-making by improving disease diagnosis. Significantly, fuzzy classification is employed in image processing, laboratory data, and determining the severity of signs and symptoms as well as the degree of sickness. The fuzzy clustering method may become a useful tool in medical diagnostics [31, 32].

9.4   Disease Diagnosis in Healthcare Sector

This section is dedicated to a number of approaches applied to FL in the diagnosis process of diseases like kidney disease, lung disease, Parkinson's disease, breast cancer, etc.

9.4.1 Kidney disease

Ahmed *et al.* [33] designed an FES to diagnose the status of a kidney. Here, status indicates the good or bad condition of the kidney. Weight, age, serum creatine, blood sugar level, intake of alcohol, and blood pressure are the input variables considered by the proposed system. The proposed approach is simple and can be used by patients. However, some of the major factors are not considered in this approach, such as water consumption level and electrolyte disorder. The accuracy of the system is high and it can be used by experts in diagnosing kidney conditions.

The authors in [34] designed a fuzzy-rule-based system for diagnosing kidney-related diseases. Symptoms are treated as features and each symptom is assigned a fuzzy value. Three physicians are asked for the suspected disease and a summary report is generated for the disease. A decision fuzzy set is obtained by fuzzy inference. Seven cases of kidney disease were identified with 63% certainty of kidney stone disease, and that of renal tubular was 15%; other cases are at different levels. One of the remarkable conclusions of this study is that the designed fuzzy expert system (FES) was fully compatible with the diagnosis of physicians.

9.4.2 Breast cancer

Breast cancer has become a common disease all over the world. In breast cancer, the cells in the breast grow rapidly. The type of cancer depends on which cells turn into cancer. This section presents some of the research done using FL to diagnose breast cancer.

Nilashi *et al.* [35] designed a rule-based FL system to diagnose breast cancer. Classification is merged with rule-based methods to classify the disease of breast cancer. Age of patient, BIRADS assessment, density, margin, and shape obtained from the mammographic dataset are the input variables used in the process of diagnosis of breast cancer. Non-incremental mining method has been applied to design the diagnosis system. Designing an incremental learning system is considered a future work of the proposed system. The system accomplished an accuracy of 94.4%, which is considered a good prediction accuracy in the diagnosis process.

Levashenko and Zaitseva [36] used fuzzy decision tree (FDT) method for the diagnosis of breast cancer. A novel method has been proposed for improving the prediction of diseases like breast cancer. Gynecological history, tumor, and heredity are used as input variables. The fuzzy nature of medical data promoted the use of FL to identify hidden patterns for improved diagnosis.

Gayathri and Sumathi [37] applied the FIS method for determining the risk of breast cancer. A number of features considered for a diagnosis of breast cancer are cell size, mitosis, the thickness of clump, cell shape, marginal adhesion, etc. Numbers of features were taken as input as the rate of breast cancer is high. This approach is concluded as it can be applied in the healthcare sector in the diagnosis process with a lesser number of attributes.

9.4.3 Diabetes

In diabetes, blood sugar levels rise rapidly due to the inability of the body to utilize it [38]. It happens when the body does not create enough insulin or does not respond effectively to it. If diabetes is not managed or treated appropriately, serious and long-term health problems might develop, such as heart difficulties, lung illnesses, liver and skin concerns, nerve loss, and so on. With the growing number of diabetes patients, early diagnosis is becoming increasingly important.

Lee and Wang [39] designed an expert system that is based on FL to help in diabetic decision-making. It is a five-layer ontology model. Ontology is a data management concept that is mostly used to construct strong links between variables. To represent the diabetes information received from data, the ontology of fuzzy diabetes is created. The five tiers of the ontology structure are the knowledge layer, group relation and group domain layer, and personal relation and personal domain layer. Over the knowledge area, the concept creation process builds concepts. The relation creation mechanism establishes the relationships between these notions. A semantic decision support agent contains the mechanism for generating ontology for a group and personal as well as a semantic fuzzy decision-making mechanism. The model's accuracy changes with the change in age characteristic, with 91.2%, 90.3%, 85.9%, 81.7%, and 77.3% for little old, little young, more or less young, very young, and very young, respectively.

By combining optimization approaches to FL to diagnose diabetic patients, Reddy and Khare [40] developed a better optimal algorithm. They created a fuzzy-rule-based forecasting algorithm, which firefly-BAT (FFBAT) improved. The algorithm is subjected to minimizing features, which makes the search space easier to navigate while maintaining precision. For this, the locality preserving projections method is applied. The data is then fuzzified, the inference engine generates rules, and the rule-based fuzzy logic does defuzzification for classification purposes. The last stage, which involves optimization, improves the system. To get the best outcomes, the firefly and BAT algorithms are combined. The complete dataset is separated into two parts: a training part that is utilized for developing rules in the fuzzy engine, and a test part that is used to verify the system's correctness.

Niswati *et al.* [41] designed an expert system that is based on FL to diagnose diabetes at an early stage. Blood pressure level, the concentration of plasma glucose, body mass index, diabetic pedigree function, and pregnancy state of a person are the factors considered. Mamdani's FIS is used to handle this data. The FIS is made up of five steps: domain issue, fuzzification, fuzzy rule development, defuzzification, and assessment. The rules in the system are created in such a way that they help in the diagnostic process by taking into account all of the input factors and their impact on the outcome. This approach was proven to be 96% accurate, with 48 of 50 instances accurately identified, matching the doctor's diagnosis. Validation testing, which checked all of the system's capabilities, questionnaire testing, which indicated the user experience, and application testing were all part of the testing process. Prediction of diabetes is also necessary to avoid the more harm that diabetes can inflict if it is not treated properly.

Rajeswari *et al.* [42] suggested a fuzzy-based association categorization approach to achieve this goal. The dataset from Pima Indian Diabetes is utilized, and a few of the pre-diabetic symptoms are used as attributes, along with some outliers. The proposed system deals with vulnerable risk marginal values, which may be handled using FLs due to their ambiguity. The association rules are used to detect patterns among the characteristics. $S\%$ support and $C\%$ confidence are used to determine the degree of correlation between the collection of qualities. The fuzzy association rules are created by matching patterns with appropriate labels of classes. The suggested technique may be implemented in three or five linguistic phrases, with the latter being the most ideal and accurate alternative. As a consequence, the approach uses association categorization to determine which risk variables, including those with a negative diabetes result, represent outlier pre-diabetic situations.

The case-based reasoning (CBR) method examines previous experience gained via the case retrieval process. Benamina *et al.* [43] developed a diabetic diagnosis based on the notion of CBR and FL. CBR entails repurposing comparable situations and assessing a novel analogous case using the technique employed in previous experience, which necessitates the acquisition of knowledge. The k closest neighbors' algorithm is used to determine how similar the examples are. The recovery of cases is the first and most important stage in CBR. Fispro, which works as the modeling portion of the fuzzy system, creates the FIS. JColibri is used for the reasoning component, which is a supplement to the first step. FDT is combined to build the case retrieval healthier in CBR.

9.4.4 Parkinson's disease

Parkinson's disease (PD) is a neurodegenerative disorder. It occurs in phases. Medication can only delay these phases, not cure them. Nilashi *et al.* [44]

developed a novel intelligent computational technique that uses the ANFIS method to enhance diagnosis and disease control. The findings demonstrated the influence of this approach on enhancing Parkinson's diagnosis and the precision of this method in forecasting disease development, as well as assisting clinicians in detecting the condition early.

Karunanithi and Rodrigues constructed an FIS model from the PD Oxford dataset to predict Parkinson's disease intensity with eight levels [45]. All conceivable combinations of four fields with the categories healthy, PD afflicted, and middle common values between healthy and PD participants were used to create a tree structure using the four fields. The intensity is computed using the following values: 12.5 for fuzzy values in healthy and PD participants and 25 for total PD impacted values. If any of these four fall under the PD impacted values, 25 will be assigned. The value is 12.5 if any of the four field values fall into the middle group or levels that are prevalent in both healthy and PD patients.

A Takagi–Sugeno FIS based on fuzzy C-means clustering is utilized to build the model [46]. The method can also be utilized to help with PD therapy. The output of the system may be used to determine the severity or level of PD in a patient. To cope with the complication of PD identification, an FL technique based on clustering is used. The study found that FIS can be taught better using fuzzy C-means clustering as compared to subtractive clustering. The accuracy of the proposed system is 96%–97%.

9.4.5 Lung cancer

lUNG cancer, being one of the most dangerous malignant tumors that may harm people's health, has a high death rate and a low cure rate due to the lack of visible early signs. When a patient is diagnosed with lung cancer, he or she is usually in the middle or advanced stages. The importance of early lung cancer screening for successful lung cancer therapy cannot be overstated. The screening procedure for lung cancer may be considered a multi-criteria decision-making issue. However, because several elements might influence lung cancer screening selections, making the appropriate decision is unstructured and difficult. Furthermore, these variables are frequently ambiguous and difficult to quantify accurately. A hesitant fuzzy set is a valuable tool for dealing with ambiguous and uncertain data, and it has a higher application in measuring it. To answer the lung cancer screening challenge, Liao *et al.*

[48] offered a framework that employs the double normalization-based multi-aggregation technique. The fuzzy Delphi technique is used to determine the key parameters for lung cancer screening.

Yilmaz *et al.* [47] devised a method for lowering the risk of death due to lung cancer by assessing the risk and the influence of stress on disease severity. This method might provide a high-accuracy early diagnosis for patients who are at risk of developing cancer. In lung cancer microarray gene expression datasets, the authors in [49] provided a tailored similarity metric for attribute selection utilizing the fuzzy rough fast reduction technique. The Shannon-entropy-based information gain filter is used to reduce dimensionality. Using current similarity metrics on a classifier that is designed using random forest, the suggested technique is tested on several datasets. It is more accurate than other similar approaches, accessible in the literature. Experimental findings of the suggested approach indicate improved accuracy of classification.

9.4.6 Thyroid

Thyroid illness is quite common all around the world, and ignoring it can lead to irreversible consequences. When endocrinologists are unavailable, Biyouki *et al.* [50] created a novel computer technique to detect thyroid disease. To build fuzzy rules based on various inputs, the authors combined the Gaussian membership function with the Sugeno inference technique.

Shariati and Haghighi [51] devised a method for detecting thyroid disorders. The results of the ANFIS approach were compared to those of other methods such as support vector machines and artificial neural networks. To conclude, the authors stated that when compared to previous approaches, their methodology improved correct diagnosis by 1.2%.

9.4.7 Skin disease

Putra and Munir [52] used an expert system to develop an application for skin condition diagnostics in children using FIS. In order to create an FIS to detect measles, German measles, and chickenpox, this system used 21 fuzzy rules based on scholarly academic articles and professionals' experience. The C# programming language was used to create the application. The built-in FIS was created using fuzzy variables as inputs, such as the temperature of the body, rash, and other symptoms associated with each skin illness.

9.4.8 Alzheimer's disease

Because Alzheimer's disease progresses slowly and in phases, Krashenyi *et al.* [53] used FIS to categorize Alzheimer's to develop the process of diagnosing the disease. Their technique is famous for allocating membership to a patient in one of the following three classifications based on MRI features: normal, MCI, or Alzheimer's disease.

9.4.9 Prostate cancer

As per the report of WHO, prostate cancer is the second most frequent malignancy among males, claiming half a million deaths each year throughout the world. When prostate cancer is organ-confined, decisive therapy, such as radical prostatectomy, is more successful. The goal of the article in [54] is to see how well certain FESs perform in classifying patients with confined or non-confined cancer. The developed technique is based on fuzzy set theory to cope with the inherent ambiguity regarding the factors used to forecast the cancer stage. A genetic algorithm was used to adjust the fuzzy rules and membership functions of an FES. As a consequence, the used method achieved improved accuracy while taking into consideration some linked studies.

9.4.10 Pneumonia disease

Pneumonia is the most frequent and widespread killer disease of the respiratory system, and it is difficult to identify since the symptoms are so similar. In [55], FES has been proposed to diagnose this. This is done to assist general practitioners and patients in making decisions and distinguishing between chronic bronchitis, tuberculosis, asthma, embolism, and lung cancer. This system went through four stages: definition, design, implementation, and testing. The sensitivity of the proposed system is 97%, specificity 85%, and accuracy 93% in diagnosing the pneumonia disease.

9.4.11 Anesthesia monitoring

Lowe [56] created the SENTINEL system, which might detect flaws and help physicians diagnose anesthetic patients. It analyzes physiological data in real time to detect pathogenic occurrences during anesthesia. Furthermore, fuzzy trend templates, a fuzzy pattern matching approach, were used to find vaguely stated trends in numerous physiological data streams. The

SENTINEL system combined the experience of numerous expert anesthetists and gained more than 90% of sensitivity and specificity accuracy for identifying seven frequent or dangerous disorders that can occur in the course of anesthesia.

Lowe and Harrison [57] created a method for diagnosing malignant hyperpyrexia (MH), an uncommon medical disease. It is based on FL. To identify variations in the patterns of indicators, rule-based diagnoses were used in this investigation. MH was recognized 9 minutes before the anesthesia in an offline assessment of the technique. These studies present the way an expert system may help to aid and improve anesthetists' performance in the medical field.

To assess the anesthetic depth, Esmaeili *et al.* [58] devised a fuzzy-rule-based technique that combines essential elements of an electroencephalogram (EEG). The information was divided into four categories: awake, light anesthesia, surgical anesthesia, and isoelectric anesthesia. The membership functions were created using statistical analysis of the given EEG features. Using training data, an ANFIS was employed to identify partitions based on the amount of DoA. An FIS and customized output membership functions were utilized to derive effective rules for this system. By simplifying the reciprocal information interchange between the human expert and the system, the study's main objective was to increase both the understanding of the results and the performance of the system.

9.4.12 Heart disease

Many ways of predicting heart disease have been presented using various methodologies. Anooj [59] used the Mamdani FIS and weighted fuzzy rules to create the system for identifying heart disease in the patient. This method has a 62.3% accuracy rate. Bashir *et al.* [60] created a heart disease prediction ensemble classifier approach that incorporated five non-homogeneous classifiers: naive Bayes, Gini index decision tree, information gain decision tree, memory-based learner, and support vector machine. The accuracy of this method was found to be 88.52%. Long *et al.* [61] created a system to diagnose heart disease, which is based on FL of interval type-2 and a rough-set-based attribute reduction (IT2FLS).

Ali-Mehdi [62] created a 94% accurate FES. The V.A. Medical Centre, Long Beach, and Cleveland Clinic Foundation databases were used in this study. They used MATLAB software to implement the project, as well as 13 input attributes and 44 rules. Baihaqi *et al.* [63] created a data mining approach that has an accuracy of 81.82%. They employed C4.5, CART, and RIPPER for the rule basis.

Polat *et al.* [64] suggested a high-accuracy artificial immune identification system and a fuzzy weighted pre-processing method for heart disease detection. A confusion matrix was used to show the system's results in this study. A fuzzy discrete hidden model was implemented by Uguz *et al.* [65]. The performance of the proposed approach outperforms that of artificial neural networks and HMM-based classification systems in this research.

Muthukaruppan [66] introduced an FES that is based on particle swarm optimization for detecting heart disease. Because the dataset contains a large number of attributes, they employed a decision tree to extract the appropriate input attributes. They have included the entire attribute information, including ranges and annotations, in this publication. Zhi *et al.* [67] built an FES for diagnosis purposes and used data mining approaches to minimize the number of characteristics. The authors utilized six input characteristics with a wide member function definition. In MATLAB software, Sikchi-Sikchi [68] created an FES using Visual Studio and a mix of MATLAB to create a graphical user interface.

Sudhakar−Manimekalai [69] proposed an approach for detecting cardiac diseases. They used a novel way of calculating the risk factor, using the FL toolkit and SQL. The centroid approach is used to defuzzify in this study.

9.5 Conclusion

Prior research on the use of fuzzy approaches in disease diagnosis has been debated in this review paper. The primary purpose of this study is to determine the impact of fuzzy approaches and their occurrence on enhancing the process of diagnosis and reducing misdiagnosed mistakes. The most striking discovery of this study is that, in addition to improving illness diagnosis, using the fuzzy technique may also enable early disease detection, which can help to avoid the advancement of complicated diseases. As a result, we expect that this evaluation will serve as a foundation for future research, as well as their classification, depending on the influence of the fuzzy approach used. We have included a number of studies that have used FL to diagnose diseases such as kidney disease, lung cancer, and heart disease.

Acknowledgement

The authors sincerely thank and appreciate the editors and reviewers for their time and valuable comments.

References

[1] R. Seising and V. Sanz, "From hard science and computing to soft science and computing – An introductory survey", Soft Computing in Humanities and Social Sciences, vol. 273, pp. 3-26, 2012.

[2] L. Zadeh, "Outline of a new approach to the analysis of complex systems and decision processes", IEEE Trans. Syst. Man Cybern. vol. 3, no. 1, pp. 28-44, 1973.

[3] A. Bhatia, V. Mago and R. Singh, "Use of soft computing techniques in medical decision making: A survey", In Advances in Computing, Communications and Informatics (ICACCI, 2014 International Conference, IEEE. 2014. p. 1131-7, Sept 2014.

[4] L. Zadeh, "Fuzzy logic, neural networks, and soft computing", Commun ACM, vol. 37, no. 3, pp.77-84, 1994.

[5] Guru99.com, [Online]. Available: https://www.guru99.com/what-is-fuzzy-logic.html, 2019. Accessed 4 July 2019.

[6] L. Zadeh, "Making computers think like people [fuzzy set theory]", IEEE Spectrum, vol. 21, no. 8, pp. 26-32, 1984.

[7] W. Pedrycz and F. Gomide, "An Introduction to Fuzzy Sets: Analysis and Design", Cambridge, Massachutes: MIT Press; 1998.

[8] B. Meunier and L. Zadeh, "Fuzzy logic and its applications", France: Addison-Wesley, pp.112-120, 1995.

[9] B. Uzun and E. Kıral, "Application of Markov chains-fuzzy states to gold price" Proc. Computer Science, vol. 120, pp. 365-371, 2017.

[10] L. Zadeh, "Fuzzy logic—a personal perspective", Fuzzy Sets and Systems, vol. 281, pp. 4-20, 2015.

[11] G. Gürsel, "Healthcare, uncertainty, and fuzzy logic", Digital Medicine, vol.2, pp.101-112, 2016.

[12] R. Seising, "Views on Fuzzy Sets and Systems from Different Perspectives: Philosophy and Logic", Criticisms and Application, Springer, vol.20, pp.34-45, 2009.

[13] E. Cox, O. Hagan, R. Table and M. Hagen, "The Fuzzy Systems Handbook with Cdrom", Orlando, FL, USA: Academic Press, Inc.; 1998.

[14] W. Pedrycz and F. Gomide, "An Introduction to Fuzzy Sets: Analysis and Design", Cambridge, Massachutes: MIT Press; 1998.

[15] F. M. McNeill and E. Thro, "Fuzzy Logic: A Practical Approach", Cambridge, Massachutes. Academic Press Professional, 1994.

[16] V. Duraisamy, N. Devarajan, D. Somasundareswari and S. N. Sivanandam, "Comparative study of membership functions for design of fuzzy logic fault diagnosis system for single phase induction motor", Academic open internet Journal, vol. 13, no. 6 pp. 1-6, 2004.

[17] G. J. Klir and T. Folger, "Fuzzy Sets, Uncertainty, and Information", Hemel Hampstead: Prentice-Hall; 1988.

[18] P. Dadone, "Design Optimization of Fuzzy Logic Systems", Doctoral Dissertation, Virginia Polytechnic Institute and State University; 2001.

[19] G. Tiruneh, A. Robinson Fayek and V. Sumati, "Neuro-fuzzy systems in construction engineering and management research", Automation in construction, Vol.119, pp.1-23, 2020.

[20] I. Persi Pamela, P. Gayathri and N. Jaisankar, "A fuzzy optimization technique for the prediction of coronary heart disease using decision tree", Int. J. Eng. Techno, vol.53, pp 101-110, 2013.

[21] S. S. Godil, M. S. Shamim, S. A. Enam and U. Qidwai, "Fuzzy logic: A "simple" solution for complexities in neurosciences", Surgical Neurology International, vol. 2, no. 24, 2011.

[22] K. Sadegh-Zadeh, "Fuzzy Health, Illness, and Disease," Journal of Medicine and Philosophy, vol.2, pp. 605-638, 2000.

[23] A. V. S. Kumar, "Fuzzy Expert Systems for Disease Diagnosis," IGI Global2014.

[24] R. A. Elisabeth, "Fuzzy and Rough Techniques in Medical Diagnosis and Medication," Springer 2007.

[25] B. E. Dixon, M. L. Kasting, S. Wilson, A. Kulkarni, G. D. Zimet and S. M. "Downs, Health care providers' perceptions of use and influence of clinical decision support reminders: qualitative study following a randomized trial to improve HPV vaccination rates," BMC Medical Informatics and Decision Making, vol.17, no.119, 2017.

[26] B. Malmir, M. Amini and S. I. Chang, "A medical decision support system for disease diagnosis under uncertainty," Expert Systems with Applications, vol. 88, pp. 95-108,2017.

[27] N. H. Phuong and V. Kreinovich, "Fuzzy logic and its applications in medicine, International journal of medical informatics," vol. 62 pp. 165-173, 2001.

[28] M. Blej and M. Aziz, "Comparison of Mamdani-Type and Sugeno-Type Fuzzy Inference Systems for Fuzzy Real Time Scheduling," International Journal of Applied Engineering Research, vol. 11, pp. 11071-11075, 2016.

[29] J. Singla, "Comparative study of Mamdani-type and Sugeno-type fuzzy inference systems for diagnosis of diabetes," 2015 International Conference on Advances in Computer Engineering and Applications, pp. 517-522, 2015.

[30] M. F. Ganji and M. S. Abadeh, "A fuzzy classification system based on Ant Colony Optimization for diabetes disease diagnosis," Expert Systems with Applications, vol. 38 pp.14650-14659, 2011.

[31] S. Shen, W. A. Sandham, M. H. Granat, M. F. Dempsey and J. Patterson, "Fuzzy clustering based applications to medical image segmentation," Proceedings of the 25th Annual International Conference of the IEEE Engineering in Medicine and Biology Society (IEEE Cat. No.03CH37439), pp. 747-750 Vol.741, 2003.

[32] M. Mahfouf, M. F. Abbod and D. A. Linken, "A survey of fuzzy logic monitoring and control utilisation in medicine," Artificial Intelligence in Medicine, vol. 21, pp. 27-42, 2001.

[33] S. Ahmed, M. T. Kabir, N. T. Mahmood and R. M. Rahman, "Diagnosis of kidney disease using fuzzy expert system, The 8th International Conference on Software," Knowledge, Information Management and Applications (SKIMA 2014), 2014, pp. 1-8.

[34] F. F. Jahantigh, B. Malmir and B. A. Avilaq, "A computer-aided diagnostic system for kidney disease", In Kidney Research and Clinical Practice, Vol, 36, pp. 29-28, 2017.

[35] M. Nilashi, O. Ibrahim, H. Ahmadi and L. Shahmoradi, "A knowledge-based system for breast cancer classification using fuzzy logic method," Telematics and Informatics, 34 (2017) 133-144.

[36] V. Levashenko and E. Zaitseva, "Fuzzy Decision Trees in medical decision Making Support System," 2012 Federated Conference on Computer Science and Information Systems (FedCSIS), 2012, pp. 213-219.

[37] B. M. Gayathri and C. P. Sumathi, "Mamdani fuzzy inference system for breast cancer risk detection," 2015 IEEE International Conference on Computational Intelligence and Computing Research (ICCIC), 2015, pp. 1-6.

[38] H. Thakkar, V. Shah, H. Yagnik and M. Shah, "Comparative anatomization of data mining and fuzzy logic techniques used in diabetes prognosis", in Clinical eHealth, Vol. 4, pp. 12-23, 2021.

[39] L. Chang-Shing and W. Mei-Hui. "A fuzzy expert system for diabetes decision support application," IEEE Trans Systems Man Cybernetics, Part B (Cybernetics). 2011;41(1):139-153.

[40] G. T. Reddy and N. Khare, "FFBAT-optimized rule based fuzzy logic classifier for diabetes," Int J Eng Res Afr. 2016;24:137-152.

[41] Z. Niswati, F. A. Mustika and A. Paramita, 2018. "Fuzzy logic implementation for diagnosis of Diabetes Mellitus disease at Puskesmas in East Jakarta," IOP Conf. Series: Journal of Physics: Conf. Series 1114.

[42] A. M. Rajeswari, M. S. Sidhika, M. Kalaivani and C. Deisy, (2018). "Prediction of Prediabetes using Fuzzy Logic based Association Classification," 2018 Second International Conference on Inventive Communication and Computational Technologies (ICICCT). doi: 10.1109/icicct.2018.8473159.

[43] M. Benamina, B. Atmani and S. Benbelkacem, "Diabetes diagnosis by case-based reasoning and fuzzy logic," Int J Interactive Multimedia Artificial Intelligence. 2018;5(3):72-80.

[44] M. Nilashi, O. Ibrahim and A. Ahani, "Accuracy Improvement for Predicting Parkinson's Disease Progression," Sci. Rep, 6 (2016).

[45] D. Karunanithi and P. Rodrigues, (2019) "A Fuzzy Rule-Based Diagnosis of Parkinson's Disease," In: Pandian D., Fernando X., Baig Z., Shi F. (eds) Proceedings of the International Conference on ISMAC in Computational Vision and Bio-Engineering 2018 (ISMAC-CVB). ISMAC 2018. Lecture Notes in Computational Vision and Biomechanics, vol 30. Springer, Cham. https://doi.org/10.1007/978-3-030-00665-5_116

[46] A. Chakraborty, A. Chakraborty and B. Mukherjee, "Detection of Parkinson's Disease Using Fuzzy Inference System", Intelligent Systems Technologies and Applications, Advances in Intelligent Systems and Computing, pp. 79-90, 2016.

[47] A. Yilmaz, S. Ari and U. Kocabicak, "Risk analysis of lung cancer and effects of stress level on cancer risk through neuro-fuzzy model," Computer methods and programs in biomedicine, 137 (2016) 35-46.

[48] H. Liao, Y. Long, M. Tang, D. Streimikiene and B. Lev, "Early lung cancer screening using double normalization-based multiaggregation (DNMA) and Delphi methods with hesitant fuzzy information," In computers and Industrial Engineering, 136, 453-463, 2019.

[49] C. Arunkumar and S. Ramakrishnan, "Attribute selection using fuzzy roughset based customized similarity measure for lung cancer microarray gene expression data," In Future Computing and Informatics Journal 3 (2018) 131-142.

[50] S. A. Biyouki, I. B. Turksen and M. H. F. Zarandi, "Fuzzy rule-based expert system for diagnosis of thyroid disease," 2015 IEEE Conference on Computational Intelligence in Bioinformatics and Computational Biology (CIBCB), 2015, pp. 1-7.

[51] S. Shariati and M. M. Haghighi, "Comparison of anfis Neural Network with several other ANNs and Support Vector Machine for diagnosing hepatitis and thyroid diseases," 2010 International Conference on Computer Information Systems and Industrial Management Applications (CISIM), 2010, pp. 596-599.

[52] A. A. Putra and R. Munir, "Implementation of fuzzy inference system in children skin disease diagnosis application," 2015 International Conference on Electrical Engineering and Informatics (ICEEI), 2015, pp. 365-370.

[53] I. Krashenyi, A. Popov, J. Ramirez and J. M. Gorriz, "Application of fuzzy logic for Alzheimer's disease diagnosis," 2015 Signal Processing Symposium (SPSympo)Poland, 2015

[54] M. J. P. Castanho, F. Hernandes, A. M. De Ré, S. Rautenberg and A. Billis, "Fuzzy expert system for predicting pathological stage of prostate cancer", In Expert Systems with Applications, 40 (2013), 466-470.

[55] L. A. Arani, F. Sadoughi and M. Langarizadeh, "An Expert System to Diagnose Pneumonia Using Fuzzy Logic", In Acta Informatica Med. 27 (2), 2019, 96-102.

[56] A. Lowe, "Evidential inference for fault diagnosis," University of Auckland: Auckland, 1998.

[57] A. Lowe and M. J. Harrison, "Computer-enhanced diagnosis of malignant hyperpyrexia" Anaes Intens Care 1999; 27(1): 41.

[58] V. Esmaeili, A. Assareh, Shamsollahi, M. H. Moradi and N. M. Arefian. "Estimating the depth of anesthesia using fuzzy soft computation applied to EEG features," Intell Data Anal. 2008; 12(4): 393-[407]

[59] P. K. Anooj, "Clinical decision support system: risk level prediction of heart disease using weighted fuzzy rules," J. King Saud Univ. Comput. Inf. Sci. 24(1), 27-40 (2012)

[60] S. Bashir, U. Qamar, F. H. Khan and M. Y. Javed, "MV5: a clinical decision support framework for heart disease prediction using majority vote based classifier ensemble," Arab. J. Sci. Eng. 39(11), 7771-7783 (2014)

[61] N. C. Long, P. Meesad and H. Unger, "A highly accurate firefly based algorithm for heart disease prediction," Expert Syst. Appl. 42(21), 8221-8231 (2015)

[62] A. Adeli and M. Neshat, "A fuzzy expert system for heart disease diagnosis," International Multi Conferences of Engineering and Computer Scientist (IMCECS2010), Hong Kong, pp. 1-6, 2010.

[63] N. A. Setiawan, I. Ardiyanto and W. M. Baihaqi, "Rule extraction for fuzzy expert system to diagnose coronary artery disease," International Conference on Information Systems and Electrical Engineering (ICITISEE), Indonesia, pp. 136-141, 2016.

[64] K. Polat, S. Güne and S. Tosun, "Diagnosis of heart disease using artificial immune recognition system and fuzzy weighted pre-processing," Pattern Recognition, vol. 39, no. 11, pp. 2186-2193, 2006.

[65] H. Uguz, A. Arslan, R. Sarac and O. I. Turkoglu, "Detection of heart valve diseases by using fuzzy discrete hidden markov model," Expert Systems with Applications, vol. 34, no. 4, pp. 2799-2811, 2008.

[66] S. Muthukaruppan, "A hybrid particle swarm optimization based fuzzy expert system for the diagnosis of coronary artery disease," Expert Systems with Applications, vol. 39, no. 14, pp.11657-11665, 2012.

[67] X. DeZhi, P. K. Butt and K. K. Oad, "A fuzzy rule based approach to predict risk level of heart disease," Global Journal of Computer Science and Technology, vol. 14, no. 3, pp. 17-22, 2014.

[68] S. S. Sikchi and S. Sikchi, "Design of fuzzy expert system for diagnosis of cardiac diseases," International Journal of Medical Science and Public Health, vol. 2, no. 1, pp. 55-61, 2013.

[69] K. Sudhakar and M. Manimekalai, "A novel methodology for diagnosing the heart disease using fuzzy database," International Journal of Research in Engineering and Technology, vol. 4, no. 10, pp. 84-89, 2015.

10

Novel Pythagorean Fuzzy Entropy-distance Measures using MCDM in the Selection of Face Masks

Mansi Bhatia, H. D. Arora, and Anjali Naithani

Department of Mathematics, Amity Institute of Applied Sciences, Amity University, India

Abstract

Wearing a face mask can help reduce the spread of infection and contamination from airborne harmful germs. The requirement to wear a face mask is perhaps one of the most noticeable lifestyle changes brought on by the COVID-19 pandemic. COVID-19 transmission can be slowed down by wearing a mask, especially while in close contact with others. Choosing the best face mask is a cumbersome task from the available alternatives in India. Several multi-criteria decision-making (MCDM) techniques and approaches have been suggested to choose the optimally probable options. The purpose of this article is to deliver an entropy–distance measure for Pythagorean fuzzy sets. To validate these measures, some of the properties were also proved. A multi-criteria decision-making approach is used to rank and hence select the best face mask for wearing. The proposed research allows the ranking of face masks based on specified criteria in a Pythagorean fuzzy environment to aid in the selection process. The results suggest that the proposed model provides a realistic way to select the best mask in the pool of considered brands. A case study on the selection process and its experimental results using Pythagorean fuzzy sets are discussed.

Keywords: COVID-19, distance measure, entropy, multi-criteria decision-making, Pythagorean fuzzy sets, TOPSIS.

10.1 Introduction

Mask has been in use for a long time for covering the face, usually for protection against germs. Until 2019, masks were worn usually by medical practitioners or by a small fraction of people to reduce the risk of alimentary diseases caused due to pollution. But 2019 drastically changed the importance of wearing a mask due to the entry of an infectious disease caused by a virus. The virus causes a disease named COVID-19, according to researchers, which spreads through the droplets generated by a person because of sneezing, coughing, or even exhaling [1]. The droplets catalyzed by the person are too heavy; so they fall on the ground [2]. The virus thus reaches the person's body and causes severe issues to the person infected. More than 200 countries have reported 2.8 billion cases out of which 5.4 million were unable to survive [3] (data taken on January 03, 2022). To date, there is no confirmed medicine to cure the disease although many vaccines have been invented and they claim to reduce the risk of losing lives and increasing the chance of a person's survival due to the formation of antibodies in the body, which can fight if the virus comes in contact. The vaccines can protect a person, but it does not claim that a person no longer needs to follow the necessary precautions against this disease [4]. One of the precautionary measures is wearing a face mask. Masks can reduce the risk of getting an infection, and, therefore, choosing the best face mask is very important [5]. Due to the wide variety of options available in the market, all claiming to be the best, it becomes difficult to decide which mask to purchase.

Decision-making is a process of choosing between a set of alternatives and arriving at a conclusion for a better outcome. When multiple criteria hold importance, making a decision can become tedious. Zadeh [6] addressed the uncertainty involved in making a distance, but with the passage of time, it was found to be inadequate. Atanassov [7, 8] suggested an intuitionistic fuzzy set (IFS), which also included non-membership (NM) function "$\varrho(x)$" along with "$\kappa(x)$" satisfying $0 \leq \kappa(x) + \varrho(x) \leq 1$ and $\kappa(x) + \varrho(x) + \lambda(x) = 1$, where $\lambda(x)$ is the hesitancy function. Later, it was found that there were sets that were not satisfying the above condition; hence, a need to revise the sets was felt. Yager [9–11] extended IFS to a Pythagorean fuzzy set (PFS) satisfying $\kappa^2(x) + \varrho^2(x) + \lambda^2(x) = 1$. Researchers have formulated many measures based on IFS and PFS using the properties of exponential, trigonometric, and many more functions. Similarity and divergence measures proposed by researchers have proved to be very helpful in a wide variety of fields. Both the concepts are opposite of each other where the former deals with the closeness of two objects and the latter talks about the distance between them. Peng [12] discussed some results on PFSs. Augustine [13] suggested both distance and similarity measures and supported them with numerical examples for better understanding. Further, Dutta and Goala [14] proposed a distance measure that was based on the concept of IFS. On the other hand, Chen [15] discussed a distance measure based on TOPSIS. Mahanta and Panda [16] have also suggested a divergence measure and applied it to real-life problems. Ohlan [17] proposed a weighted PF exponential distance measure depicting the effectiveness with the help of a numerical example. Gao *et al.* [18] proposed a divergence measure related to PFS and used the measure in pattern recognition.

Multi-attribute decision-making (MADM) is a method that helps in choosing the best option out of a set of alternatives available. Hwang and Yoon [19] gave the theory in 1981 with further developments by Hwang [20] in 1993. It has been applied in different areas of statistics, mathematics, psychology, environment, etc. Chen [21] proposed IFS fuzzy TOPSIS. Lin *et al.* [22] suggested applying MCDM based on IFS, thus making it easy for decision-makers to conclude. Zhang and Xu [23] extended TOPSIS to MCDM with PFSs. Zulqarnain *et al.* [24] applied TOPSIS for DM. Verma and Maheshwari [25] proposed a divergence measure and applied it in MCDM with a fuzzy environment. Agrawal [26] proposed a divergence measure with order and degree as α and β, respectively, under PFSs. Sangwan and Kaur [27] did a review on the TOPSIS approach from real to fuzzy sets. Ye and Chen [28] proposed TOPSIS under PFSs for the selection of the best cotton fabrics based on their properties. Many researchers combine more than one MCDM method to have an effective comparison and study the problem more effectively. Bączkiewicz [29] applied TOPSIS, VIKOR, PROMETHEE II, and COMET to show the usefulness of MCDM in E-commerce. Vakilipour *et al.* [30] also did a comparison between different MCDM methods to evaluate the quality of life at different levels. Bhatia *et al.* [31] proposed a sine divergence measure and applied TOPSIS for selecting the best property out of the suitable options available. While making a decision, sometimes, there are many attributes that are to be considered among the options available. Each attribute has its weightage. Many a time, weights are chosen hypothetically by researchers, but finding appropriate weights can give better results. There are a lot of MCDM methods to find out weights like SAW, AHP entropy, etc.

Entropy is defined as the degree of uncertainty or randomness by Shannon [32]. Fuzzy entropy was proposed by Zadeh [6]. It has been used by many researchers for finding out weights. Entropy is inversely proportional to the weight of the attribute; the larger the entropy, the smaller will be the attribute weight. Applying entropy to find out the weight in the TOPSIS algorithm helps in giving better results. Zheng *et al.* [33] proposed an entropy–distance measure based on PFS and applied TOPSIS to it. Hussain [34] discussed fuzzy entropy based on PFS and applied it in MCDM. Chen [35] studied the effect of entropy weights on TOPSIS for a large number of data. Further, Gandotra *et al.* [36] suggested a Pythagorean entropy measure and proved that its entropy can determine the information in PFS. Arora *et al.* [36] discussed the logarithmic entropy distance measure and applied it to post-COVID implications.

The rest of the chapter is divided into sections as follows. Section 10.2 is for preliminaries and Section 10.3 is for the proposed divergence measure. Section 10.4 is dedicated to numerical calculations and Section 10.5 is for applications through TOPSIS. Section 10.6 concludes the chapter.

10.2 Preliminaries

Definition 1 ([6]). Let $\mathcal{H}$ be defined as a fuzzy set in a non-empty set X. Then it is represented as

$$\mathcal{H} = \{<x, \kappa_{\mathcal{H}}(x)|x \in X\}, \tag{10.1}$$

where $\kappa_{\mathcal{H}}(x)$: X $\rightarrow$ [0, 1] is the degree of membership of an element x in set X.

Definition 2 ([7]). Let $\mathcal{H}$ be an intuitionistic fuzzy set in X. Then we can define

$$\mathcal{H} = \{<x, \ \kappa_{\mathcal{H}}(x), \varrho_{\mathcal{H}}(x)|\forall x \in X\}, \tag{10.2}$$

where $\kappa_{\mathcal{H}}(x)$: X $\rightarrow$ [0, 1] and $\varrho_{\mathcal{H}}(x)$: X $\rightarrow$ [0, 1].

$\kappa_{\mathcal{H}}(x)$ is the degree of belongingness and $\varrho_{\mathcal{H}}(x)$ is the degree of non-belongingness such that $0 \leq \kappa_{\mathcal{H}}(x) + \varrho_{\mathcal{H}}(x) \leq 1$.

Definition 3 ([9]). A PFS is given as

$$\mathcal{H} = \{<x, \kappa_{\mathcal{H}}(x), \varrho_{\mathcal{H}}(x)\} \ \forall X \in. \tag{10.3}$$

And $\kappa_{\mathcal{H}}(x)$: X $\rightarrow$ [0, 1] and $\varrho_{\mathcal{H}}(x)$: X $\rightarrow$ [0, 1], where $\kappa_{\mathcal{H}}(x)$ is the degree of membership and $\varrho_{\mathcal{H}}(x)$ is the degree of non-membership (NM) such that $0 \leq \kappa_{\mathcal{H}}^2(x) + \varrho_{\mathcal{H}}^2(x) \leq 1$ and $\lambda_{\mathcal{H}}^2(x) = 1 - \kappa_{\mathcal{H}}^2(x) - \varrho_{\mathcal{H}}^2(x)$ where $\lambda_{\mathcal{H}}(x)$ is called hesitancy or indeterminacy of PFS $\mathcal{H}$.

PFSs are a generalization of IFS and they allow better representation of uncertain information than IFS. As shown in Figure 10.1, PFSs contain IFS along with values not satisfied by IFS.

Definition 4 ([32]).

Let X be a discrete random variable where $X = \{x_1, x_2, \ldots, x_n\}$ with probability $P(x_1), P(x_2), \ldots, P(x_n)$. Then the entropy function is defined as

$$H(X) = -\sum_{i=1}^{n} P(x_i) \log P(x_i).$$ (10.4)

Definition 5 ([21]). Let $\mathbb{G} = \{\mathfrak{g}_1, \mathfrak{g}_2, \mathfrak{g}_3\}$ and $\mathbb{H} = \{\mathfrak{h}_1, \mathfrak{h}_2, \mathfrak{h}_3\}$ be two fuzzy sets. Then the Euclidean distance is defined as

$$D(\mathbb{G}, \mathbb{H}) = \sqrt{\frac{1}{3}\{(\mathfrak{h}_1 - \mathfrak{g}_1)^2 + (\mathfrak{h}_2 - \mathfrak{g}_2)^2 + (\mathfrak{h}_3 - \mathfrak{g}_3)^2}.$$ (10.5)

Definition 6 ([16]). Let $\mathbb{G} = \{\mathbb{G}_A(x_i), \mathbb{G}_B(x_i)\}$ and $\mathbb{H} = \{\mathbb{H}_A(x_i), \mathbb{H}_B(x_i)\}$ be two PFSs such that $X = \{x_1, x_2, \ldots, x_n\}$. Then the

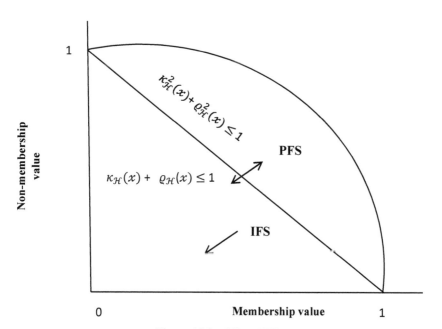

Figure 10.1   IFS vs. PFS.

distance between the two of them is given by

$$D(\mathbb{G}, \mathbb{H}) = \frac{1}{n} \sum_{i=1}^{n} \frac{\left|\mathbb{G}_\mathbb{A}^2(x_i) - \mathbb{H}_\mathbb{A}^2(x_i)\right| + \left|\mathbb{G}_\mathbb{B}^2(x_i) - \mathbb{H}_\mathbb{B}^2(x_i)\right|}{\mathbb{G}_\mathbb{A}^2(x_i) + \mathbb{H}_\mathbb{A}^2(x_i) + \mathbb{G}_\mathbb{B}^2(x_i) + \mathbb{H}_\mathbb{B}^2(x_i)} \qquad (10.6)$$

10.3 Proposed Distance Measure

In this article, we have proposed a distance measure between two PFSs as PFSs are stronger in representing uncertainty than IFS; hence, they prove themselves as a better tool for expressing uncertainty when making decisions. Now we will discuss the proposed distance measure between two PFSs $\mathbb{G}$ and $\mathbb{H}$ and discuss the properties satisfied by it in detail.

Definition 3.1. Let $X = \{x_1, x_2, \ldots, x_n\}$ be a non-empty set and $\mathbb{G}$ and $\mathbb{H}$ be the two PFSs, where $\mathbb{G}$ and $\mathbb{H}$ can be defined as

$$\mathbb{G} = \{ x, \kappa_\mathbb{G}(x_i), \varrho_\mathbb{G}(x_i)\} and \ \mathbb{H} = \{ x, \kappa_\mathcal{H}(x_i), \varrho_\mathcal{H}(x_i)\} \forall x_i \in X.$$

Then the distance measured between them can be given as

$$S^1(\mathbb{G}, \mathbb{H}) = 1 - \frac{\sqrt{2}}{n} \sum_{i=1}^{n} \sin \frac{\pi}{2} - \frac{\pi}{4}$$
$$(\left|\kappa_\mathbb{G}^2(x_i) - \kappa_\mathbb{H}^2(x_i)\right| \vee \left|\varrho_\mathbb{G}^2(x_i) - \varrho_\mathbb{H}^2(x_i)\right|$$
$$\vee \left|\lambda_\mathbb{G}^2(x_i) - \lambda_\mathbb{H}^2(x_i)\right| + 1) \qquad (10.7)$$

$$S^2(\mathbb{G}, \mathbb{H}) = 1 - \frac{\sqrt{2}}{n} \sum_{i=1}^{n} w_i \sin[\frac{\pi}{2} - \frac{\pi}{4}$$
$$(\left|\kappa_\mathbb{G}^2(x_i) - \kappa_\mathbb{H}^2(x_i)\right| \vee \left|\varrho_\mathbb{G}^2(x_i) - \varrho_\mathbb{H}^2(x_i)\right|$$
$$\vee \left|\lambda_\mathbb{G}^2(x_i) - \lambda_\mathbb{H}^2(x_i)\right| + 1)] \qquad (10.8)$$

where $\lambda_\mathbb{G}^2(x_i) = 1 - \kappa_\mathbb{G}^2(x_i) - \varrho_\mathbb{G}^2(x_i)$ and $\lambda_\mathbb{H}^2(x_i) = 1 - \kappa_\mathbb{H}^2(x_i) - \varrho_\mathbb{H}^2(x_i)$, and $w = (w_1, w_2, \ldots, w_n)^T$ denotes the weight of variables $x_i \ \forall \ i \in 1, 2, \ldots, n$ and $w_m \in [0, 1]$ where $m \in \{1, 2, \ldots, n\}$ and $\sum_{m=1}^{n} w_m = 1$.

For this, $S^1(\mathbb{G}, \mathbb{H})$ and $S^2(\mathbb{G}, \mathbb{H})$ $\mathbb{G}$ and $\mathbb{H}$ in X the sine distance measure will have to satisfy the following four properties:

1. $0 \leq S^1(\mathbb{G},\mathbb{H}), S^2(\mathbb{G},\mathbb{H}) \leq 1$.

2. $S^1(\mathbb{G},\mathbb{H})$, $S^2(\mathbb{G},\mathbb{H}) = 0 \Leftrightarrow \mathbb{G} = \mathbb{H}$.
3. $S^1(\mathbb{G},\mathbb{H}) = S^1(\mathbb{H},\mathbb{G})$ and $S^2(\mathbb{G},\mathbb{H}) = S^2(\mathbb{H},\mathbb{G})$.
4. (a) If $\mathbb{G}$ is a PFS in X and $\mathbb{G} \subseteq \mathbb{H} \subseteq \mathbb{I}$, then $S^1(\mathbb{G}, \mathbb{H}) \leq S^1(\mathbb{G},\mathbb{I})$ and $S^1(\mathbb{H}, \mathbb{I}) \leq S^1(\mathbb{G},\mathbb{I})$.

(b) If $\mathbb{G}$ is a PFS in X and $\mathbb{G} \subseteq \mathbb{H} \subseteq \mathbb{I}$, then $S^2(\mathbb{G}, \mathbb{H}) \leq S^2(\mathbb{G},\mathbb{I})$ and $S^2(\mathbb{H}, \mathbb{I}) \leq S^2(\mathbb{G},\mathbb{I})$.

Proof:

1. $0 \leq S^1(\mathbb{G}, \mathbb{H}) \leq 1$:
Since $0 \leq \left|\kappa_{\mathbb{G}}^2(x_i) - \kappa_{\mathbb{H}}^2(x_i)\right| \leq 1$, $0 \leq \left|\varrho_{\mathbb{G}}^2(x) - \varrho_{\mathbb{H}}^2(x)\right| \leq 1$, $0 \leq \left|\lambda_{\mathbb{G}}^2(x) - \lambda_{\mathbb{H}}^2(x)\right| \leq 1$

$\Rightarrow 0 \leq \left|\kappa_{\mathbb{G}}^2(x) - \kappa_{\mathbb{H}}^2(x)\right| \vee \left|\varrho_{\mathbb{G}}^2(x) - \varrho_{\mathbb{H}}^2(x)\right| \vee \left|\lambda_{\mathbb{G}}^2(x) - \lambda_{\mathbb{H}}^2(x)\right| + 1 \leq 2$

Multiplying $-\frac{\pi}{4}$ on both the sides
$\Rightarrow 0 \geq \left|\kappa_{\mathbb{G}}^2(x) - \kappa_{\mathbb{H}}^2(x)\right| \vee \left|\varrho_{\mathbb{G}}^2(x) - \varrho_{\mathbb{H}}^2(x)\right| \vee \left|\lambda_{\mathbb{G}}^2(x) - \lambda_{\mathbb{H}}^2(x)\right| + 1 \geq -\frac{\pi}{2}$

Adding $\frac{\pi}{2}$ on both the sides,
$\Rightarrow 0 \leq \left|\kappa_{\mathbb{G}}^2(x) - \kappa_{\mathbb{H}}^2(x)\right| \vee \left|\varrho_{\mathbb{G}}^2(x) - \varrho_{\mathbb{H}}^2(x)\right| \vee \left|\lambda_{\mathbb{G}}^2(x) - \lambda_{\mathbb{H}}^2(x)\right| + 1 \leq \frac{\pi}{2}$

$\because$ Sine lies between [0, 1]. Thus,

$$\Rightarrow 0 \leq \sin[\frac{\pi}{2} - \frac{\pi}{4} \; \left|\kappa_{\mathbb{G}}^2(x_i) - \kappa_{\mathbb{H}}^2(x_i)\right| \vee \left|\varrho_{\mathbb{G}}^2(x_i) - \varrho_{\mathbb{H}}^2(x_i)\right|$$
$$\vee \left|\lambda_{\mathbb{G}}^2(x_i) - \lambda_{\mathbb{H}}^2(x_i)\right| + 1] \leq 1$$

$$\Rightarrow 0 \leq 1 - \frac{\sqrt{2}}{n} \sum_{i=1}^{n} \sin[\frac{\pi}{2} - \frac{\pi}{4}[\left|\kappa_{\mathbb{G}}^2(x_i) - \kappa_{\mathbb{H}}^2(x_i)\right| \vee \left|\varrho_{\mathbb{G}}^2(x_i) - \varrho_{\mathbb{H}}^2(x_i)\right|$$
$$\vee \left|\lambda_{\mathbb{G}}^2(x_i) - \lambda_{\mathbb{H}}^2(x_i)\right| + 1]] \leq 1$$
$$\Rightarrow 0 \leq S^1(\mathbb{G}, \mathbb{H}) \leq 1.$$

A similar proof can be made for $S^2(\mathbb{G}, \mathbb{H})$.

2. $S^1(\mathbb{G}, \mathbb{H}) = 0 \Leftrightarrow \mathbb{G} = \mathbb{H}$

$$\Leftrightarrow 1 - \frac{\sqrt{2}}{n} \sum_{i=1}^{n} \sin[\frac{\pi}{2} - \frac{\pi}{4} \; \left|\kappa_{\mathbb{G}}^2(x_i) - \kappa_{\mathbb{H}}^2(x_i)\right| \vee \left|\varrho_{\mathbb{G}}^2(x_i) - \varrho_{\mathbb{H}}^2(x_i)\right|$$
$$\vee \left|\lambda_{\mathbb{G}}^2(x_i) - \lambda_{\mathbb{H}}^2(x_i)\right| + 1] = 0$$

$$\Leftrightarrow \sin[\frac{\pi}{2} - \frac{\pi}{4} |\kappa_{\mathbb{G}}^2(x) - \kappa_{\mathbb{H}}^2(x)| \vee |\varrho_{\mathbb{G}}^2(x) - \varrho_{\mathbb{H}}^2(x)|$$

$$\vee |\lambda_{\mathbb{G}}^2(x) - \lambda_{\mathbb{H}}^2(x)| + 1] = \frac{1}{\sqrt{2}}$$

$$\Leftrightarrow \frac{\pi}{2} - \frac{\pi}{4} |\kappa_{\mathbb{G}}^2(x) - \kappa_{\mathbb{H}}^2(x)| \vee |\varrho_{\mathbb{G}}^2(x) - \varrho_{\mathbb{H}}^2(x)|$$

$$\vee |\lambda_{\mathbb{G}}^2(x) - \lambda_{\mathbb{H}}^2(x)| + 1 = \frac{\pi}{4}$$

$$\Leftrightarrow -\frac{\pi}{4} |\kappa_{\mathbb{G}}^2(x) - \kappa_{\mathbb{H}}^2(x)| \vee |\varrho_{\mathbb{G}}^2(x) - \varrho_{\mathbb{H}}^2(x)|$$

$$\vee |\lambda_{\mathbb{G}}^2(x) - \lambda_{\mathbb{H}}^2(x)| + 1 = -\frac{\pi}{4}$$

$$\Leftrightarrow |\kappa_{\mathbb{G}}^2(x) - \kappa_{\mathbb{H}}^2(x)| \vee |\varrho_{\mathbb{G}}^2(x) - \varrho_{\mathbb{H}}^2(x)| \vee |\lambda_{\mathbb{G}}^2(x) - \lambda_{\mathbb{H}}^2(x)| = 0$$

$$\Leftrightarrow \mathbb{G}=\mathbb{H}.$$

A similar proof can be done for $S^2(\mathbb{G}, \mathbb{H})$.

3. $S^1(\mathbb{G},\mathbb{H}) = S^1(\mathbb{H}, \mathbb{G})$ and $S^2(\mathbb{G},\mathbb{H}) = S^2(\mathbb{H}, \mathbb{G})$.

The proof is obvious.

4. If $\mathbb{G}$ is a PFS in X and $\mathbb{G} \subseteq \mathbb{H} \subseteq \mathbb{I}$, then $S^1(\mathbb{G}, \mathbb{H}) \leq S^1(\mathbb{G}, \mathbb{I})$ and $S^1(\mathbb{H}, \mathbb{I}) \leq S^1(\mathbb{G}, \mathbb{I})$ and $S^2(\mathbb{G}, \mathbb{H}) \leq S^2(\mathbb{G}, \mathbb{I})$ and $S^2(\mathbb{H}, \mathbb{I}) \leq S^2(\mathbb{G}, \mathbb{I})$.

For $S^1(\mathbb{G}, \mathbb{H})$, if $\mathbb{G} \subseteq \mathbb{H} \subseteq \mathbb{I}$, then $\forall x_i \in X, 0 \leq \kappa_{\mathbb{G}}(x_i) \leq \kappa_{\mathbb{H}}(x_i) \leq \kappa_{\mathbb{I}}(x_i) \leq 1$ and $1 \geq \varrho_{\mathbb{G}}(x_i) \geq \varrho_{\mathbb{H}}(x_i) \geq \varrho_{\mathbb{I}}(x_i) \geq 0$.

$\Rightarrow 0 \leq \kappa_{\mathbb{G}}^2(x_i) \leq \kappa_{\mathbb{H}}^2(x_i) \leq \kappa_{\mathbb{I}}^2(x_i) \leq 1$ and $1 \geq \varrho_{\mathbb{G}}^2(x_i) \geq \varrho_{\mathbb{H}}^2(x_i) \geq \varrho_{\mathbb{I}}^2(x_i) \geq 0$

$\Rightarrow |\kappa_{\mathbb{G}}^2(x_i) - \kappa_{\mathbb{H}}^2(x_i)| \leq |\kappa_{\mathbb{G}}^2(x_i) - \kappa_{\mathbb{I}}^2(x_i)|, |\kappa_{\mathbb{H}}^2(x_i) - \kappa_{\mathbb{I}}^2(x_i)| \leq |\kappa_{\mathbb{G}}^2(x_i) - \kappa_{\mathbb{I}}^2(x_i)|$ &

$\Rightarrow |\varrho_{\mathbb{G}}^2(x_i) - \varrho_{\mathbb{H}}^2(x_i)| \leq |\varrho_{\mathbb{G}}^2(x_i) - \varrho_{\mathbb{I}}^2(x_i)|, |\varrho_{\mathbb{H}}^2(x_i) - \varrho_{\mathbb{I}}^2(x_i)| \leq |\varrho_{\mathbb{G}}^2(x_i) - \varrho_{\mathbb{I}}^2(x_i)|$

$\Rightarrow |\lambda_{\mathbb{G}}^2(x_i) - \lambda_{\mathbb{H}}^2(x_i)| \leq |\lambda_{\mathbb{G}}^2(x_i) - \lambda_{\mathbb{I}}^2(x_i)|, |\lambda_{\mathbb{H}}^2(x_i) - \lambda_{\mathbb{I}}^2(x_i)| \leq |\lambda_{\mathbb{G}}^2(x_i) - \lambda_{\mathbb{I}}^2(x_i)|$

$\Rightarrow |\kappa_{\mathbb{G}}^2(x_i) - \kappa_{\mathbb{H}}^2(x_i)| \vee |\varrho_{\mathbb{G}}^2(x_i) - \varrho_{\mathbb{H}}^2(x_i)| \vee |\lambda_{\mathbb{G}}^2(x_i) - \lambda_{\mathbb{H}}^2(x_i)| \leq |\kappa_{\mathbb{G}}^2(x_i) - \kappa_{\mathbb{I}}^2(x_i)| \vee |\varrho_{\mathbb{G}}^2(x_i) - \varrho_{\mathbb{I}}^2(x_i)| \vee |\lambda_{\mathbb{G}}^2(x_i) - \lambda_{\mathbb{I}}^2(x_i)|,$

$\Rightarrow |\kappa_{\mathbb{H}}^2(x_i) - \kappa_{\mathbb{I}}^2(x_i)| \vee |\varrho_{\mathbb{H}}^2(x_i) - \varrho_{\mathbb{I}}^2(x_i)| \vee |\lambda_{\mathbb{H}}^2(x_i) - \lambda_{\mathbb{I}}^2(x_i)| \leq |\kappa_{\mathbb{G}}^2(x_i) - \kappa_{\mathbb{I}}^2(x_i)| \vee |\varrho_{\mathbb{G}}^2(x_i) - \varrho_{\mathbb{I}}^2(x_i)| \vee |\lambda_{\mathbb{G}}^2(x_i) - \lambda_{\mathbb{I}}^2(x_i)|$

$$\Rightarrow \sin[\frac{\pi}{2} - \frac{\pi}{4}(|\kappa_{\mathbb{G}}^2(x_i) - \kappa_{\mathbb{H}}^2(x_i)| \vee |\varrho_{\mathbb{G}}^2(x_i) - \varrho_{\mathbb{H}}^2(x_i)|$$

$$\vee |\lambda_{\mathbb{G}}^2(x_i) - \lambda_{\mathbb{H}}^2(x_i)| + 1)] \geq \sin[\frac{\pi}{2} - \frac{\pi}{4}(|\kappa_{\mathbb{G}}^2(x_i) - \kappa_{\mathbb{H}}^2(x_i)|$$

$$\vee |\varrho_{\mathbb{G}}^2(x_i) - \varrho_{\mathbb{H}}^2(x_i)| \vee |\lambda_{\mathbb{G}}^2(x_i) - \lambda_{\mathbb{H}}^2(x_i)| + 1)]$$

$$\Rightarrow \sin[\frac{\pi}{2} - (|\kappa_{\mathbb{G}}^2(x_i) - \kappa_{\mathbb{H}}^2(x_i)| \vee |\varrho_{\mathbb{G}}^2(x_i) - \varrho_{\mathbb{H}}^2(x_i)|$$

$$\vee |\lambda_{\mathbb{G}}^2(x_i) - \lambda_{\mathbb{H}}^2(x_i)| + 1)] \geq \sin[\frac{\pi}{2} - \frac{\pi}{4}(|\kappa_{\mathbb{G}}^2(x_i) - \kappa_{\mathbb{H}}^2(x_i)|$$

$$\vee |\varrho_{\mathbb{G}}^2(x_i) - \varrho_{\mathbb{H}}^2(x_i)| \vee |\lambda_{\mathbb{G}}^2(x_i) - \lambda_{\mathbb{H}}^2(x_i)| + 1)]$$

$$\Rightarrow 1 - \frac{\sqrt{2}}{n}\sin[\frac{\pi}{2} - \frac{\pi}{4}(|\kappa_{\mathbb{G}}^2(x_i) - \kappa_{\mathbb{H}}^2(x_i)| \vee |\varrho_{\mathbb{G}}^2(x_i) - \varrho_{\mathbb{H}}^2(x_i)|$$

$$\vee |\lambda_{\mathbb{G}}^2(x_i) - \lambda_{\mathbb{H}}^2(x_i)| + 1)] \leq 1 - \frac{\sqrt{2}}{n}\sin[\frac{\pi}{2} - \frac{\pi}{4}(|\kappa_{\mathbb{G}}^2(x_i) - \kappa_{\mathbb{I}}^2(x_i)|$$

$$\vee |\varrho_{\mathbb{G}}^2(x_i) - \varrho_{\mathbb{I}}^2(x_i)| \vee |\lambda_{\mathbb{G}}^2(x_i) - \lambda_{\mathbb{I}}^2(x_i)| + 1)]$$

$$\Rightarrow 1 - \frac{\sqrt{2}}{n}\sin[\frac{\pi}{2} - \frac{\pi}{4}(|\kappa_{\mathbb{H}}^2(x_i) - \kappa_{\mathbb{I}}^2(x_i)| \vee |\varrho_{\mathbb{H}}^2(x_i) - \varrho_{\mathbb{I}}^2(x_i)|$$

$$\vee |\lambda_{\mathbb{H}}^2(x_i) - \lambda_{\mathbb{I}}^2(x_i)| + 1)] \leq 1 - \frac{\sqrt{2}}{n}\sin[\frac{\pi}{2} - \frac{\pi}{4}(|\kappa_{\mathbb{G}}^2(x_i) - \kappa_{\mathbb{I}}^2(x_i)|$$

$$\vee |\varrho_{\mathbb{G}}^2(x_i) - \varrho_{\mathbb{I}}^2(x_i)| \vee |\lambda_{\mathbb{G}}^2(x_i) - \lambda_{\mathbb{I}}^2(x_i)| + 1)$$

$$\Rightarrow S^1(\mathbb{G}, \mathbb{H}) \leq S^1(\mathbb{G}, \mathbb{I}) \text{ and } S^1(\mathbb{H}, \mathbb{I}) \leq S^1(\mathbb{G}, \mathbb{I}).$$

Similarly, $S^2(\mathbb{G}, \mathbb{H}) \leq S^2(\mathbb{G}, \mathbb{I})$ and $S^2(\mathbb{H}, \mathbb{I}) \leq S^2(\mathbb{G}, \mathbb{I})$.

10.4 Numerical Illustration

In this section, we will numerically verify the measure proposed with the help of the example given below:

$$\mathbb{G} = \{\langle x_1, 0.7, 0.1 \rangle, \langle x_2, 0.3, 0.7 \rangle, \langle x_3, 0.6, 0.2 \rangle\},$$
$$\mathbb{H} = \{\langle x_1, 0.7, 0.2 \rangle, \langle x_2, 0.8, 0.2 \rangle, \langle x_3, 0.5, 0.2 \rangle\},$$
$$\mathbb{I} = \{\langle x_1, 0.8, 0.3 \rangle, \langle x_2, 0.9, 0.1 \rangle, \langle x_3, 0.8, 0.2 \rangle\}.$$

Using eqn (10.7), we get

$$S^1(\mathbb{G}, \mathbb{H}) = 1 - \frac{\sqrt{2}}{6}[\sin(\frac{\pi}{2} - \frac{\pi}{4}(|0.7^2 - 0.7^2| \vee |0.1^2 - 0.2^2|$$

$\vee |0.5 - 0.47| + 1) + \sin(\frac{\pi}{2} - \frac{\pi}{4}(|0.3^2 - 0.8^2| \vee |0.7^2 - 0.2^2|$

$\vee |0.42 - 0.32|) + \sin(\frac{\pi}{2} - \frac{\pi}{4}(|0.6^2 - 0.5^2| \vee |0.2^2 - 0.2^2|$

$\vee |0.6 - 0.71|) + 1)].$

$S^1(\mathbb{G}, \mathbb{H}) = 1 - \frac{\sqrt{2}}{6} [\sin(\frac{\pi}{2} - \frac{\pi}{4}(|0| \vee |0.03| \vee |0.03| + 1)$

$+ \sin(\frac{\pi}{2} - \frac{\pi}{4}(|0.55| \vee |0.45| \vee |0.1| + 1)$

$+ \sin(\frac{\pi}{2} - \frac{\pi}{4}(|0.11| \vee |0| \vee |0.11|) + 1)].$

$S^1(\mathbb{G},\mathbb{H}) = 1 - \frac{\sqrt{2}}{6} (0.690251 + 0.346117 + 0.643456).$

$S^1(\mathbb{G},\mathbb{H}) = 0.6040.$

$S^1(\mathbb{H}, \mathbb{I}) = 1 - \frac{\sqrt{2}}{6} [\sin(\frac{\pi}{2} - \frac{\pi}{4}[(|0.7^2 - 0.8^2| \vee |0.2^2 - 0.3^2|$

$\vee |0.47 - 0.27| + 1) + \sin(\frac{\pi}{2} - \frac{\pi}{4}(|0.8^2 - 0.9^2| \vee |0.2^2 - 0.1^2|$

$\vee |0.32 - 0.18|) + \sin(\frac{\pi}{2} - \frac{\pi}{4}(|0.5^2 - 0.8^2| \vee |0.2^2 - 0.2^2|$

$\vee |0.71 - 0.32|) + 1)].$

$S^1(\mathbb{H}, \mathbb{I}) = 1 - \frac{\sqrt{2}}{6} [\sin(\frac{\pi}{2} - \frac{\pi}{4}[(|0.15| \vee |0.05| \vee |0.2| + 1)$

$+ \sin(\frac{\pi}{2} - \frac{\pi}{4} (|0.17| \vee |0.03| \vee |0.14| + 1)$

$+ \sin(\frac{\pi}{2} - \frac{\pi}{4}(|0.39| \vee |0| \vee |0.39|) + 1)].$

$S^1(\mathbb{H}, \mathbb{I}) = 1 - \frac{\sqrt{2}}{6} (0.587785 + 0.606682 + 0.460974).$

$S^1(\mathbb{H}, \mathbb{I}) = 0.6098.$

$S^1(\mathbb{G}, \mathbb{I}) = 1 - \frac{\sqrt{2}}{6} [\sin(\frac{\pi}{2} - \frac{\pi}{4}[(|0.7^2 - 0.8^2| \vee |0.1^2 - 0.3^2|$

$\vee |0.5 - 0.27| + 1) + \sin(\frac{\pi}{2} - \frac{\pi}{4}(|0.3^2 - 0.9^2| \vee |0.7^2 - 0.1^2|$

$\vee |0.42 - 0.18|) + \sin(\frac{\pi}{2} - \frac{\pi}{4}(|0.6^2 - 0.8^2| \vee |0.2^2 - 0.2^2|$

$\vee |0.6 - 0.32|) + 1)].$

$$S^1(\mathbb{G},\ \mathbb{I}) = 1 - \frac{\sqrt{2}}{6}\ [\ \sin(\frac{\pi}{2} - \frac{\pi}{4}[(|0.15|\ \vee\ |0.08|\ \vee\ |0.23| + 1)$$

$$+ \sin(\frac{\pi}{2} - \frac{\pi}{4}\ (|0.72|\ \vee\ |0.48|\ \vee\ |0.24| + 1)$$

$$+ \sin(\frac{\pi}{2} - \frac{\pi}{4}(|0.28|\ \vee\ |0|\ \vee\ |0.28|) + 1)].$$

$$S^1(\mathbb{G},\mathbb{I}) = 1 - \frac{\sqrt{2}}{6}\ (0.568562\ +\ 0.218143\ + 0.535827).$$

$$S^1(\mathbb{G},\mathbb{I}) = 0.6882.$$

Similarly, the values of $S^2(\mathbb{G},\ \mathbb{H})$, $S^2(\mathbb{H},\ \mathbb{I})$, and $S^2(\mathbb{G},\ \mathbb{I})$ are 0.8800, 0.8652, and 0.9088, respectively.

Numerical rationale: From the computations, following observations can be drawn:

1. $0 \le S^i\ (\mathbb{G},\ \mathbb{H}) \le 1;\ \ i = 1,\ 2.$
2. $S^i\ (\mathbb{G},\mathbb{H}) = 1 \Leftrightarrow \mathbb{G} = \mathbb{H};\ \ i = 1,\ 2.$
3. $S^i\ (\mathbb{G},\ \mathbb{H}) = S^i\ (\mathbb{H},\ \mathbb{G})\ ;\ i = 1,\ 2.$
4. $S^i\ (\mathbb{G},\ \mathbb{H}) \le S^i\ (\mathbb{H},\ \mathbb{I})$ and $S^i\ (\mathbb{G},\mathbb{I}) \le S^i\ (\mathbb{H},\ \mathbb{I})\ ;\ \ i = 1,\ 2.$

Thus, from the above calculations, we can say that the proposed measure is validated.

10.5 Application Through TOPSIS

The TOPSIS algorithm is one of the easiest and most commonly used MCDM methods. It gives a quick and easy solution to complex situations involving a lot of criteria. Also, it gives a qualitative and quantitative view of the problem.

10.5.1 Stepwise explanation of the TOPSIS algorithm:

Step 1: A decision matrix is constructed using PFSs.

Let $\mathbb{A} = \{\mathbb{A}_1,\ \mathbb{A}_2, \ldots, \mathbb{A}_n\}$ be a set with n alternatives and the decision maker has to choose the best one from $\mathbb{A}$ based on a set $\mathbb{C} = \{\mathbb{C}_1,\ \mathbb{C}_2, \ldots, \mathbb{C}_m\}$ containing m criteria. A $n \times m$ matrix $\mathbb{V} = \left[v_{ij}^k\right]_{n \times m}$ is formed, where $v_{ij} = (\kappa_{ij},\ \varrho_{ij})$ is structured. Here, κ_{ij} and ϱ_{ij} are the membership and NM grades of the alternatives $\mathbb{A}_i$ satisfying the criteria $\mathbb{C}_i$. The PFSs index $\lambda_{ij} = \sqrt{1 - \kappa_{ij}^2 - \varrho_{ij}^2}$ denotes the decision maker's indeterminacy grade of the alternative $\mathbb{A}_i$ with respect to criteria$\mathbb{C}_i$. A PFS from fuzzy sets is constructed in this step.

Step 2: Normalize the fuzzy decision matrix.

The matrix so formed has to be normalized; to do so, the membership and non-membership values for cost criteria are interchanged, whereas for benefit criteria, it remains interchanged.

Step 3: Calculate the weights for the criteria.

The weights w_j used here are calculated by the entropy method.

Here, $w_j = \frac{d_{ij}}{1-d_{ij}}$, $d_{ij} = 1 - e_{ij}$ and $e_{ij} = -\frac{\sum N_{ij}\ ln(N_{ij})}{ln\ (m\ )}$, m = number of criteria.

Step 4: Calculate the fuzzy positive ideal solution and fuzzy negative ideal solution.

FPIS and FNIS are calculated by using eqn (10.9) and (10.10):

$$
\begin{aligned}
\mathbb{A}^{k+} &= \left\{ r_1^{k+}, r_2^{k+}, \ldots, r_n^{k+} \right\} \\
&= \left\{ \left(\max_m (r_{mn}^k)/n \in M \right), \left(\left(\min_m (r_{mn}^k)/n \in N \right) \right) \right\}
\end{aligned}
\tag{10.9}
$$

and

$$
\begin{aligned}
\mathbb{A}^{k-} &= \left\{ r_1^{k-}, r_2^{k-}, \ldots, r_n^{k-} \right\} \\
&= \left\{ \left(\min_m (r_{mn}^k)/n \in M \right), \left(\left(\max_m (r_{mn}^k)/n \in N \right) \right) \right\}.
\end{aligned}
\tag{10.10}
$$

Here, M and N represent benefit and cost criteria, respectively.

Step 5: Determine the distance of each alternative.

The distance of each alternative is calculated by using the divergence formula as in eqn (10.11) and (10.12):

$$
S^2(\mathbb{A}, \mathbb{A}_i^+) = \mathbb{D}\left(\mathbb{A}, \mathbb{A}_i^+ \right) = 1 - \frac{\sqrt{2}}{n} \sum_{i=1}^n w_i \sin\left[\frac{\pi}{2} - \frac{\pi}{4}(|\kappa_\mathbb{A}^2(x_i) - \kappa_{\mathbb{A}_i^+}^2(x_i) \right.
$$
$$
\left. | \vee |\varrho_\mathbb{A}^2(x_i) - \varrho_{\mathbb{A}_i^+}^2(x_i)| \vee |\lambda_\mathbb{A}^2(x_i) - \lambda_{\mathbb{A}_i^+}^2(x_i)| + 1) \right].
\tag{10.11}
$$

$$
S^2(\mathbb{A}, \mathbb{A}_i^-) = \mathbb{D}\left(\mathbb{A}, \mathbb{A}_i^- \right) = 1 - \frac{\sqrt{2}}{n} \sum_{i=1}^n w_i \sin\left[\frac{\pi}{2} - \frac{\pi}{4}\left(\left| \kappa_\mathbb{A}^2(x_i) - \kappa_{\mathbb{A}_i^-}^2(x_i) \right| \right. \right.
$$
$$
\left. \left. \vee \left| \varrho_\mathbb{A}^2(x_i) - \varrho_{\mathbb{A}_i^-}^2(x_i) \right| \vee \left| \lambda_\mathbb{A}^2(x_i) - \lambda_{\mathbb{A}_i^-}^2(x_i) \right| + 1) \right].
\tag{10.12}
$$

Step 6: Compute the closeness coefficient using eqn (10.13).

$$\mathfrak{N}_i = \frac{\mathbb{D}\left(\mathbb{A}, \mathbb{A}_i^-\right)}{\mathbb{D}\left(\mathbb{A}, \mathbb{A}_i^+\right) + \mathbb{D}\left(\mathbb{A}, \mathbb{A}_i^-\right)} = \frac{\mathcal{S}_i^-}{\mathcal{S}_i^+ + \mathcal{S}_i^-}. \qquad (10.13)$$

Step 7: Formulate the rank of the alternatives.

The rank of alternatives is calculated based on their closeness coefficient such that the highest value is the best ideal solution.

The TOPSIS algorithm is also depicted in the flowchart as shown in Figure 10.2.

A contagious disease called COVID-19 is a worldwide spreader of severely acute disease caused by SARS-CoV-2 (coronavirus) since 2019. Its first case emerged in China and since then, it has created a pandemic situation with minimal to dangerous outcomes such as loss of life. To date, there is no effective medicine for it. Although vaccines have been developed for protection against it, still precautionary measures have to be followed to safeguard oneself from it. One of the most important measures is wearing a

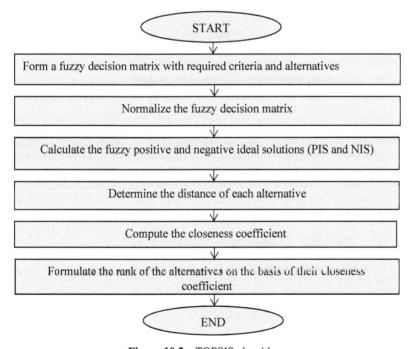

Figure 10.2 TOPSIS algorithm.

mask as the virus enters the lungs through the nostrils and mouth. Although the mask is not a new thing and people use it for other reasons also, it has gained a lot of importance because of COVID-19. As it is a part of lifestyle, people consider various factors before buying it. Some important criteria include price (C1), style (C2), color (C3), material (C4), number of layers (C5), and number of permissible washes (C6). Assume that you have six masks: Mask 1 (M1), Mask 2 (M2), Mask 3 (M3), Mask 4 (M4), Mask 5 (M5), and Mask 6 (M6). A mask requires to utilize PFSs to do a better assessment based on the above criteria. Six masks are to be evaluated by the decision maker based on six criteria as given below:

Step 1: Data in the form of PFN by decision makers is given in Table 10.1.

Step 2: Based on this data, masks are to be ranked and the best mask is to be determined. In the MADM technique, the first step is to identify benefit and cost criteria. The benefit criteria, as the name suggests, should be higher, and the cost criteria is to be kept minimum. In the above criteria, style, color, material, and permissible washes are the benefit criteria and price and number of layers are cost criteria.

Data needs to be normalized, which is done by interchanging the membership and non-membership values for cost criteria and no interchange in the case of benefit criteria as shown in Table 10.2.

Table 10.1　Data set as a decision matrix.

Mask	Price	Style	Color	Material	No. of layers	Washable
Mask 1	<0.7,0.2>	<0.6,0.3>	<0.4,0.2>	<0.5,0.5>	<0.4,0.3>	<0.9,0.2>
Mask 2	<0.8,0.3>	<0.7,0.1>	<0.35,0.25>	<0.7,0.4>	<0.6,0.4>	<0.7,0.3>
Mask 3	<0.2,0.6>	<0.3,0.2>	<0.2,0.2>	<0.4,0.3>	<0.7,0.2>	<0.4,0.4>
Mask 4	<0.1,0.5>	<0.5,0.4>	<0.3,0.1>	<0.3,0.4>	<0.6,0.2>	<0.4,0.5>
Mask 5	<0.3,0.4>	<0.9,0.2>	<0.6,0.5>	<0.3,0.7>	<0.3,0.1>	<0.7,0.3>
Mask 6	<0.4,0.6>	<0.4,0.3>	<0.1,0.8>	<0.4,0.4>	<0.5,0.2>	<0.6,0.6>

Table 10.2　Normalization of the decision matrix.

Mask	Price	Style	Color	Material	No. of layers	Washable
M1	<0.2,0.7>	<0.6,0.3>	<0.4,0.2>	<0.5,0.5>	<0.3 , 0.4>	<0.9,0.2>
M2	<0.3,0.8>	<0.7,0.1>	<0.35,0.25>	<0.7,0.4>	<0.6,0.4>	<0.7,0.3>
M3	<0.6,0.2>	<0.3,0.2>	<0.2,0.2>	<0.4,0.3>	<0.2,0.7>>	<0.4,0.4>
M4	<0.5,0.1>	<0.5,0.4>	<0.3,0.1>	<0.3,0.4>	<0.6,0.2>	<0.4,0.5>
M5	<0.4,0.3>	<0.9,0.2>	<0.6,0.5>	<0.3,0.7>	<0.1,0.3>	<0.7,0.3>
M6	<0.6,0.4>	<0.4,0.3>	<0.1,0.8>	<0.4,0.4>	<0.2,0.5>	<0.6,0.6>

Table 10.3 Positive and negative ideal solutions for each criterion.

	Price	Style	Color	Material	No. of layers	Washable
A+	<0.2,0.8>	<0.9,0.1>	<0.6,0.1>	<0.7,0.3>	<0.1,0.7>	<0.9,0.2>
A−	<0.6,0.1>	<0.3,0.4>	<0.1,0.8>	<0.3,0.7>	<0.6,0.2>	<0.4,0.6>

Table 10.4 Separation measures for fuzzy positive ideal solution.

Mask	Price	Style	Color	Material	No. of layers	Washable
M1	0.9952	0.9872	0.9877	0.9814	0.9753	0.9865
M2	0.9952	0.9890	0.9874	0.9856	0.9768	0.9749
M3	0.9964	0.9828	0.9849	0.9739	0.9730	0.9692
M4	0.9956	0.9856	0.9854	0.9739	0.9739	0.9692
M5	0.9955	0.9930	0.9910	0.9771	0.9681	0.9786
M6	0.9967	0.9842	0.9858	0.9766	0.9739	0.9749

Step 3: Fuzzy positive ideal solution (FPIS) and fuzzy negative ideal solution (FNIS) are shown in Table 10.3.

Step 4: As it is known to receive the best alternative, the geometrical distance with benefit criteria should be minimum, and with cost criteria, it should be maximum. The distance measure is calculated using eqn (10.11) and (10.12) as shown in Tables 10.4 and 10.5. The weights w_j used in the measure calculated by using entropy method are $w_1 = 0.038792$, $w_2 = 0.123076$, $w_3 = 0.113153$, $w_4 = 0.232949$, $w_5 = 0.251131$, and $w_6 = 0.240899$.

Here, $w_j = \frac{d_{ij}}{1-d_{ij}}$, $d_{ij} = 1 - e_{ij}$ and $e_{ij} = -\frac{\sum N \ln(N)}{\ln(m)}$, m = number of criteria.

Step 5: The coefficient of the relationship is calculated by using eqn (10.13) as shown in Table 10.6.

Table 10.5 Separation measures for fuzzy negative ideal solution.

Mask	Price	Style	Color	Material	No. of layers	Washable
M1	0.9973	0.9897	0.9862	0.9814	0,9744	0.9692
M2	0.9967	0.9879	0.9862	0.9771	0.9833	0.9783
M3	0.9964	0.9869	0.9838	0.9739	0.9831	0.9762
M4	0.9962	0.9913	0.9843	0.9739	0.9803	0.9794
M5	0.9961	0.9832	0.9890	0.9842	0.9670	0.9783
M6	0.9964	0.9897	0.9932	0.9766	0.9761	0.9819

Table 10.6   Correlation coefficient and ranking obtained from TOPSIS.

Mask	S_i^+	S_i^+	$\mathfrak{N}_i$	Rank
M1	5.9136	5.8985	0.4993	6
M2	5.9092	5.9098	0.5000	4
M3	5.0005	5.0005	0.5008	3
M4	5.8837	5.9055	0.5009	2
M5	5.9035	5.8981	0.4997	5
M6	5.8923	5.9142	0.5009	1

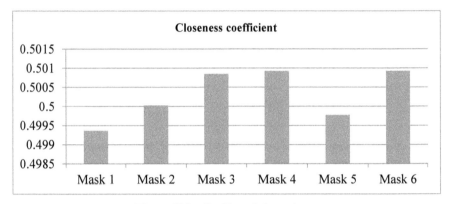

Figure 10.3   Ranking of alternatives.

Based on closeness, coefficient rank has been obtained as shown in Table 10.6, and the result is obtained as M6 ≻ M4 ≻ M3 ≻ M2 ≻ M5 ≻ M1; the same has been depicted graphically in Figure 10.3.

The result obtained shows that Mask 6 has the highest closeness coefficient, showing that it is the best alternative out of the chosen alternatives.

10.6 Conclusion

A new sine distance measure has been proposed and its axiomatic properties have also been proved. Numerical illustrations have been given to prove the validity of the measure. The TOPSIS algorithm has been formulated using the proposed distance measure and a real-life problem has been taken to prove the authenticity of the measure. Charts have been enclosed for a better understanding of the proposed example.

TOPSIS is one of the most commonly used MCDM techniques due to its simple nature. It is a convenient method for the decision makers, which helps in finding the best among a number of alternatives available. For assigning the weights in this method, many different techniques have been adopted such as the entropy method, AHP, and also hypothetical assumptions. In this chapter, we have used the entropy measure for finding the weights, which increase the credibility of the measure. The proposed measure is applied for the selection of the best face mask for COVID-19 using TOPISS. The novelties of the proposed method are that it offers an easy and objective evaluation, which gives reliable and practical results. The application of the entropy method for finding weights double ensures the validity of the proposed distance measure. This study can be applied to different situations like risk management, education, medicine, and agriculture, where complex decision-making is required. Also, other techniques have been applied for finding out weights like AHP, and a comparison between them can be done to see the difference in weights calculated and the effect (if any) they have on the measure.

Acknowledgement

The authors wholeheartedly acknowledge the appraisers and the editorial team for spending their time carefully reading the article and for providing us with valuable comments.

References

[1] A. Wawrzyniak, K. Kuczborska, A. L. Opałka, A. Będzichowska, B. Kalicki. 'The 2019 novel coronavirus (2019-nCoV) – transmission, symptoms and treatment'. Pediatria i Medycyna Rodzinna, pp C1-C5, Dec., 2019.

[2] Z. J. Cheng, J. Shan. '2019 Novel coronavirus: where we are and what we know. Infection', pp 155-163, Apr., 2020.

[3] World Health Organization Novel Coronavirus (SARS-CoV-2 virus), Jan. 2022.

[4] World Health Organization Novel Coronavirus (SARS-CoV-2 virus), Jan. 2022.

[5] World Health Organization Novel Coronavirus (SARS-CoV-2 virus), Jan. 2022

[6] L. A. Zadeh. 'Fuzzy sets, Information and Control', pp. 338–353, Jun., 1965.

[7] K. Atanassov. 'Intuitionistic fuzzy sets. Fuzzy Sets Syst.', pp. 87–96, Aug., 1986.

[8] K. Atanassov. 'More on intuitionistic fuzzy sets. Fuzzy Sets Syst.', pp. 37–46, Oct., 1989.

[9] R. Yager. 'Pythagorean fuzzy subsets', Proc. Joint IFSA World Congress and NAFIPS Annual Meeting (IFSA/NAFIPS), pp. 57–61, Canada, Jun., 2013.

[10] R. Yager. 'Pythagorean membership grades in multi criteria decision making', In: Technical report MII-3301.Machine Intelligence Institute, Iona College, New Rochelle, Mar., 2013.

[11] R. Yager. 'Pythagorean membership grades in multi criteria decision making', IEEE Trans Fuzzy Syst., pp. 958– 965, Aug., 2014.

[12] X. Peng. 'Some Results for Pythagorean Fuzzy Sets'. Wiley Online Library, IJIS Int. Journal of of Intelligent Syst.., pp 1133-1160, May., 2015.

[13] E. Augustine. 'Distance and similarity measures for Pythagorean fuzzy sets' Granul Comput., Dec., 2018.

[14] P. Dutta, S. Goala. 'Fuzzy Decision Making in Medical Diagnosis Using an Advanced Distance Measure on Intuitionistic Fuzzy Sets'. The Open Cybernetics & Systemics Journal, 2018.

[15] F. Zhou, T. Chen. 'A Novel Distance Measure for Pythagorean Fuzzy Sets and its Applications to the Technique for Order Preference by Similarity to Ideal Solutions', IJCIS Int. Journal of Comput Intelligence Syst. pp 955-969, 2019.

[16] J. Mahanta, S. Panda. 'Distance measure for Pythagorean fuzzy sets with varied applications'. Neural Comput & Applic, Aug.2021.

[17] A. Ohlan. 'Multiple attribute decision-making based on distance measure under pythagorean fuzzy environment'. IJIT Int. Journal of Information Tech., 2021.

[18] X. Gao, L. Pan, Y. Deng. 'Uncertainty Measure of Pythagorean Fuzzy Sets'. Fuzzy systems and data mining VII, pp. 183-189, Oct. 2021.

[19] C. L. Hwang, K. Yoon. 'Methods for Multiple Attribute Decision Making. In: Multiple Attribute Decision Making'. Lecture Notes in Economics and Mathematical Systems, pp. 58-191. 1981.

[20] C. L. Hwang., Y. J Lai, T. Y. Liu. 'A new approach for multiple objective decision making', Computers and Operational Research, 1993.

[21] T. Y. Chen, C. Y. Tsao.'The interval-valued fuzzy TOPSIS method and experimental analysis'. Fuzzy Sets and Systems,2008.

[22] L. Lin, X. H. Yuan, Z. Q. Xia. 'Multicriteria fuzzy decision-making methods based on intuitionistic fuzzy sets', Journal of Computer and System Sciences, 2007.

[23] X. Zhang, X. Zeshui. 'Extension of TOPSIS to Multiple Criteria Decision Making with Pythagorean Fuzzy Sets', IJIS International Journal of Intelligent Systems 2007.

[24] R. M. Zulqarnain, M. Saeed, N. Ahma, B. Ahmad, F. Dayan. 'Application of TOPSIS Method for Decision Making' IJSRMSS International Journal of Scientific Research in Mathematical and Statistical Sciences, 2020.

[25] R. Verma, S. Maheshwari. 'A new measure of divergence with its application to multi-criteria decision making under fuzzy environment'. Neural Comput & Applic, 2017.

[26] N. Agarwal. 'Parametric Directed Divergence Measure for Pythagorean Fuzzy Set and Their Applications to Multi-criteria Decision-Making'. In: Garg H. (eds) Pythagorean Fuzzy Sets.,2021.

[27] R. Sangwan, G. Kaur, 'A Review on Topsis Approach: From Real to Fuzzy Settings', AJOMCOR Asian Journal of Mathematics and Computational Research, 2021.

[28] J. Ye, T. Y. Chen. ' Selection of Cotton Fabrics Using Pythagorean Fuzzy TOPSIS Approach,' Journal of Natural Fibers, 2021.

[29] A. Bączkiewicz. 'MCDM based e-commerce consumer decision support tool', Procedia Computer Science, 2021.

[30] S. Vakilipour, A. S. Niaraki, M. Ghodousi, S. M. Choi. 'Comparison between Multi-Criteria Decision-Making Methods and Evaluating the Quality of Life at Different Spatial Levels', 2021.

[31] M. Bhatia, HD Arora, A Naithani, S Gupta. 'Distance measures of Pythagorean Fuzzy sets based on sine function in property selection under TOPSIS approach. 12th International Conference on Cloud Computing, Data Science & Engineering (Confluence) 1-7, IEEE.

[32] C. E. Shannon. 'A Mathematical Theory of Communication'. Bell System Technical Journal, 1948.

[33] Q. Han, W. Li, Y. Lu, M. Zheng, W. Quan, Y. Song. 'TOPSIS Method Based on Novel Entropy and Distance Measure for Linguistic Pythagorean Fuzzy Sets with Their Application in Multiple Attribute Decision Making'. IEEE Access. 2019.

[34] Z. Hussain, S. Abbas, S. Hussain, Z. Ali, G. Jabeen. 'Similarity measures of Pythagorean fuzzy sets with applications to pattern recognition

and multi criteria decision making with Pythagorean TOPSIS'. Journal of mechanics of continua and mathematical sciences, 2021.

[35] P. Chen. 'Effects of the entropy weight on TOPSIS', Expert Systems with Applications, 2021.

[36] S. R. Kumar, N. Gandotra. 'Novel Pythagorean Fuzzy Entropy with Application in MCDM to assess the Best Automotive Company'. 8th International Conference on Computing for Sustainable Global Development (INDIA.Com), pp. 167-171, 2021.

[37] H. D. Arora, A. Naithani. 'Effectiveness of Logarithmic Entropy measures for Pythagorean Fuzzy Sets in diseases related to post Covid implications under TOPSIS Approach'. IJISAE International Journal of Intelligent Systems and Applications in Engineering, 2021.

11

Prioritizing the Barriers of Manufacturing during COVID-19 using Fuzzy AHP

Shwetank Avikal[1], Rushali Pant[2], K. C. Nithin Kumar[3], Vimal Kumar[4], and Mangey Ram[5,6]

[1]Department of Management Studies, Graphic Era Hill University, India
[2]Department of Mechanical Engineering, Graphic Era Hill University, India
[3]Department of Mechanical Engineering, Graphic Era Deemed to be University, India
[4]Department of Information Management, Chaoyang University of Technology, Taiwan
[5]Graphic Era Deemed to be University, India
[6]Institute of Advanced Manufacturing Technologies, Peter the Great St. Petersburg Polytechnic University, Russia

Abstract

COVID-19 has wreaked havoc on the global economy, supply chains, and government, posing an unparalleled health threat. The manufacturing sector was one of the most disruptive systems in the world at the time of the COVID-19 pandemic. Most manufacturing companies have faced a lock-down situation and are focusing on the production of essential products. Furthermore, COVID-19 has altered customer behavior. The short-term and long-term effects of COVID-19 on the manufacturing sector must be evaluated to hasten recovery and build preparedness measures should another such disruption occur. The limitations affecting the construction system during this period were discussed and prioritized in this study. The ambiguous nature of human thinking makes it difficult to evaluate the qualitative parameters; hence, it is preferred to incorporate an approach that converts the variables into triangular fuzzy numbers to better represent the values of the criteria. A fuzzy analytical hierarchical procedure (FAHP) is applied to evaluate the limitation criteria in an ambiguous environment. "Growing demand for existing products" is considered the heaviest limit after "financial stagnation" and "setback in logistics services." The study results will help the manufacturing company in formulating and implementing strategies to overcome the pandemic situation.

Keywords: Barriers, COVID-19, coronavirus, fuzzy analytic hierarchy process, manufacturing.

11.1 Introduction

Coronavirus infection (COVID-19) has affected the global economy to shrink rapidly. The epidemic is causing severe disruption to the economy due to slowdowns in many sectors such as construction, education, tourism, manufacturing, and transportation industries. The government is adopting various measures such as flight shutdown, lockdown, and quarantine to resolve the situation.

The manufacturing industry is one of the most affected sectors due to the pandemic outbreak. The sector has faced supply and demand imbalances due to production shutdowns, improper inventory management, logistics services, and high risk and uncertainty. This was triggered by the closure of manufacturing operations in China and had an impact on the supply of raw materials and spare parts [7]. Since then, major manufacturing companies have faced plant shutdowns, leading to imbalances in product supply and demand. The logistics shock did not create a complete demand for products in the isolated and process industries. This collapse of the manufacturing sector is due to several limitations and is a major concern of this study.

The aim of this study is to discuss and prioritize limiting factors in the manufacturing sector during the COVID-19 period. Barriers are identified and extracted from the relevant literature. To prioritize identified factors, a fuzzy analytical hierarchical process (FAHP) is implemented. The process is capable of detecting weights of various scales. Fuzzy logic is used to include uncertainty in the priorities of decision-making. The results of this study may be beneficial to manufacturing company managers in determining a strategy to deal with an epidemic situation.

The article is divided into five sections. Section 11.1 discusses the introduction and needs of the study. Section 11.2 discusses the articles available in the literature along with the identification of the barriers. Section 11.3 outlines the research methodology adopted in this study. Section 11.4 is enlisting and discussing the results obtained from the AHP method, and Section 11.5 is having concluding remarks on the study.

11.2 Literature Review

The impact of COVID-19 on the manufacturing industry is explored and discussed by a few consultancy firms, academicians, and practitioners. This section provides a brief discussion of the relevant studies and identifies the barriers.

The author studied the impact on the manufacturing sector during and after COVID-19 [7]. To help revive the manufacturing supply chain, several countermeasures have been suggested and discussed. The impact of COVID-19 is discussed by considering the case of the automobile and aviation industries to study the supply chain resilience theory in the manufacturing and service sectors [5]. A combination of qualitative and quantitative approaches is applied to assess the strategies followed by the case industries.

Some studies have focused on assessing the impact of COVID-19 on countries' economies. Some estimated the impact of growth, manufacturing, trade, and small and medium enterprises on the Indian economy [17]. The study shows that the manufacturing sector is likely to shrink by 5.5%−20% compared to last year. Researchers also discuss the impact of COVID-19 on the manufacturing and services sector of manufacturing companies in northern Italy [16]. Extensive surveys and interviews were conducted for data collection and a four-stage crisis-management model for improved post-pandemic conditions was presented. Authors have predicted the impact of the COVID-19 pre-crisis on the UK manufacturing industry [13]. Strategies for a comprehensive overhaul of the manufacturing system after COVID-19 have been suggested.

Study of barriers to production and expectations of promoters during COVID-19 estimate [15] is based on a survey of 71 manufacturing.... industries in the Americas, Europe, Africa, and Asia. Some recommendations have been proposed and discussed to improve the situation in the manufacturing sector.

The fuzzy AHP method addressed a variety of problems in many types of research. The fuzzy AHP technique has been applied for evaluating and selecting a supplier for steel pipe supplier selection problems [22]. Others applied the fuzzy AHP method to prioritize the factors that support DevOps practices in software organizations [1]. Enablers toward a sustainable supply chain were analyzed using fuzzy AHP [18]. Fuzzy AHP and Bayesian network were applied for the risk assessment of gas explosions in coal mines [14]. Romanian manufacturing companies were evaluated using fuzzy AHP and TOPSIS [4]. The process was also applied for detecting malware in android mobiles [4]. The e-service quality in the airline industry from the customer's perspective was evaluated using the same [3]. An integrated model of life cycle sustainability assessment (LCSA), fuzzy AHP as multi-criteria decision analysis (MCDA) and building information model (BIM) was used to select sustainable material for buildings [11].

It is clear from the literature review that previous studies have discussed the impact of COVID-19 on the manufacturing sector or estimated limitations using qualitative methods. Fuzzy AHP method can be used to evaluate both qualitative and quantitative criteria. Also, it is necessary to perform criteria comparisons based on TFNs. By comparing them, weight values can be determined. In this study, a two-step approach to identifying and prioritizing barriers was adopted so that mitigation strategies could be determined.

11.2.1 Identification of Barriers in the Manufacturing Industry

To identify the conceptual content in any field, a literature review is the best methodology to follow. The barriers in the manufacturing sector during COVID-19 are identified using a literature review and enlisted with references in Table 11.1.

The coronavirus originated in the city of Wuhan in China and spread around the world. Most companies in the world rely on the Chinese manufacturing sector. The cessation of supply of raw materials and spare parts from China has increased the demand for existing products, which has not been met. Assembling lines of major manufacturing companies have been shut down due to reliance on spare parts. Demand for automobile products, accessories, and other luxury goods plummeted due to the immediate lockdown and confinement policies of the government and the closure of several factories [23].

Demand for food and pharmaceutical goods has increased. Many companies suffered sale losses and job cuts due to financial constraints [5]. Due to limitations, operation and logistics cost increased. Many designs and new product development strategies call for government-limited teamwork and procurement.

Table 11.1 List of barriers to the manufacturing sector with reference.

S. no.	Barriers	References
1	Lack of technical information and capability	[15]
2	COVID-19-related health and safety concerns	[7, 13, 15]
3	Lack of positive response from the government upon helping	[15]
4	Government imposed restrictions	[5, 7, 13, 15]
5	Increased demand for existing products	[7, 15]
6	Lack of resources and skillset	[7, 15]
7	Setbacks in logistics services	[7]
8	Time constraints	[15]
9	Financial constraints	[5, 7, 15, 17]
10	Complexity in repurposing products and infrastructure	[15]
11	Supply and demand issue	[5, 7, 13, 15]

11.3 Research Methodology

The methodology followed in this study was twofold: identify barriers in the construction area using a critical literature review and prioritize removed barriers to find the relative weight of each barrier. To this end, relevant articles were collected from the Scopus database, thoroughly reviewed, and obstacles were found. The method of priority is discussed in this section. The process chart is shown in Figure 11.1.

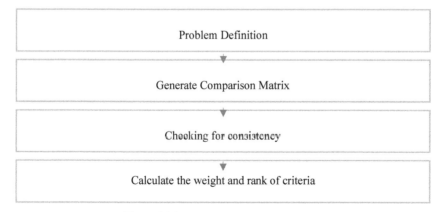

Figure 11.1 Process chart of fuzzy AHP.

11.3.1 Analytical Hierarchy Process

AHP was proposed by Satty in 1981. This is also known as the eigenvector method. AHP assigns relevant significance by pair-by-comparison of each criterion. In pair-by-pair comparisons, AHP requires real numbers to scale in the $1-9$ range. In pair-by-pair comparisons, a pair-by-pair comparison matrix should be constructed in which they compare each criterion with all other criteria. Some field experts need to design the matrix in pairs and these experts compare the standards with other criteria based on their previous knowledge and skills. Many researchers around the world have implemented AHP in their work and found it to be an appropriate tool for assessing relative importance [10]. It was found that experts use their experience and reserve knowledge to make subjective judgments [8].

11.3.2 Fuzzy Set Theory

Fuzzy set theory works with scales 0 and 1. All information about the fuzzy set is extracted from its membership function. Elements for all input values related to the membership function are mapped within the interval [0, 1]. A minimum value of 0 indicates that the element does not belong to the set. Value 1 indicates that element belongs to the whole set. A value element between 0 and 1 indicates that it has some membership. The obscure set must contain an element with a specific membership [19].

11.3.3 Computational Procedure of Fuzzy AHP

AHP has been implemented worldwide to assess the relative importance of selected criteria. Pair-wise comparison matrix *MK* should be formed with expert advice, where k refers to the number of professionals involved in the complex process of brainwashing and decision-making. Many decision-makers have accumulated knowledge of the past and appropriate judgment to predict outcomes. When solving, as the magnitude of the complications increases, it appears to be an obstacle to achieving more accurate results. Therefore, to get more accurate results, existing problem-solving methods need to be updated [9, 12].

Therefore, the AHP method does not appear to be effective in obtaining accurate results for decision-making for any real-life problem. The integration of the fuzzy set theory into AHP is more effective in solving complex decision-making problems. In the fuzzy AHP method, the decision-maker uses the fuzzy value on the actual number for pair-by-comparison. The computational process of AHP was provided to implement fuzzy AHP tools in complex problem solving [20]. For computation, the procedure of fuzzy AHP has been used as a reference [19].

11.4 Result and Discussion

The barriers discussed in the literature sections have been considered as criteria for the study as shown in Table 11.2. The importance of these criteria has been computed with the help of fuzzy AHP.

A hierarchy has been designed as shown in Figure 11.2, which shows the relationship between the final decision and the decision criteria.

These parameters are paired with each other on a fuzzy scale. A team of experts was selected to create a pair-wise comparison matrix. A total of 18 experts were asked for this and the last pair-wise comparison matrix is shown in Table 11.3.

The significance/weight of these criteria is calculated with the help of a pair-wise comparison matrix and is shown in Table 11.4. The results show that the "Increased demand for existing products" criterion is the most important criterion and has the highest weight. The second and third highest

Table 11.2 Selection of barriers as decision criteria.

S. no.	Barriers	Criteria no.
1	Lack of technical information and capability	C1
2	COVID-19-related health and safety concerns	C2
3	Lack of positive response from the government upon offering assistance	C3
4	Government-imposed restrictions	C4
5	Increased demand for existing products	C5
6	Lack of resources and skillset	C6
7	Setbacks in logistics services	C7
8	Time constraints	C8
9	Financial constraints	C9
10	Complexity in repurposing product/infrastructure	C10
11	Supply and demand issue	C11

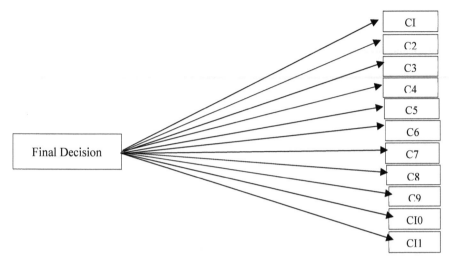

Figure 11.2   Decision hierarchy for the barriers.

Table 11.3   Pair-wise comparison matrix of selection criteria.

	C1	C2	C3	C4	C5	C6	C7	C8	C9	C10	C11
C1	1	1/3	1	1/5	1/7	1/5	1/7	1/5	1/7	1/5	1/3
C2	3	1	5	1/3	1/5	1/3	1/5	1/3	1/5	1/3	3
C3	1/1	1/5	1	1/7	1/7	1/5	1/7	1/7	1/7	1/5	1/3
C4	5	3	7	1	1/3	3	1/3	3	1/3	3	5
C5	7	5	7	3	1	5	3	5	3	5	7
C6	5	3	5	1/3	1/5	1	1/3	1/3	1/5	3	3
C7	7	5	7	3	1/3	3	1	3	1/3	3	5
C8	5	3	7	1/3	1/5	3	1/3	1	1/3	3	5
C9	7	5	7	3	1/3	5	3	3	1	5	5
C10	5	3	5	1/3	1/5	1/3	1/3	1/3	1/5	1	1
C11	3	1/3	3	1/5	1/7	1/3	1/5	1/5	1/5	1/1	1

weights calculated are 0.185814 and 0.141128 for "Financial constraints" and "Setbacks in logistics services," respectively. The minimum weight calculated is 0.015666 for "Lack of positive response from the government upon offering assistance." "Lack of technical information and capability" is of second least concern with 0.017467 followed by "Supply and demand issues" with 0.030328 weight. The consistency check for the pair-wise comparison matrix was also performed and the stability ratio value was found to be less than 0.1.

Table 11.4   Criteria weights calculated by fuzzy-AHP.

Criteria	Weight	Consistency
Lack of technical information and capability	0.017467	CI = 0.147077
		CR = 0.097402
COVID-19-related health and safety concerns	0.043131	
Lack of positive response from the government upon offering assistance	0.015666	
Government-imposed restrictions	0.109877	
Increased demand for existing products	0.24762	
Lack of resources and skillset	0.067146	
Setbacks in logistics services	0.141128	
Time constraints	0.089808	
Financial constraints	0.185814	
Complexity in repurposing product/infrastructure	0.052014	
Supply and demand issue	0.030328	

11.5 Conclusion and Future Research Avenues

The study focuses on identifying and prioritizing barriers in the manufacturing sector in the context of the COVID-19 epidemic. Limitations were identified from a critical review of the relevant literature and prioritized using a vague analytical hierarchical approach. The analysis found that "increasing demand for existing products" is a high-level lifting limit. The identification and analysis provided in this study are helpful in determining mitigation and resilience strategies to combat disruption.

Future studies may focus on risk assessment and mitigation strategies after COVID-19. The epidemic has hit small- and medium-sized industries very hard, which requires a lot of studies. In addition, studies on proposing strategies to make the productive supply chain more resilient to disruptive events are another good direction. Further studies on Industry 4.0 and similar approaches to digitization are needed to address the impact of artificial intelligence, machine learning, and smart manufacturing.

References

[1] M. A. Akbar, S. Mahmood, M. Shafiq, A. Alsanad, A. A. A. Alsanad and A. Gumaei, 'Identification and prioritization of DevOps success factors using fuzzy-AHP approach.' Soft Computing, pp. 1-25, 2020.

[2] J. M. Arif, M. F. Ab Razak, S. R. T. Mat, S. Awang, N. S. N. Ismail and A. Firdaus, 'Android mobile malware detection using fuzzy AHP.' Journal of Information Security and Applications, vol. 61, pp. 102929, 2021.

[3] M. Bakır and Ö. Atalık, 'Application of fuzzy AHP and fuzzy MAR-COS approach for the evaluation of e-service quality in the airline industry.' Decision Making: Applications in Management and Engineering, vol. 4, no. 1, pp. 127-152, 2021.

[4] A. I. Ban, O. I., Ban, V. Bogdan, D. C. S. Popa and D. Tuse, 'Performance evaluation model of Romanian manufacturing listed companies by fuzzy AHP and TOPSIS.' Technological and Economic Development of Economy, vol. 26, no. 4, pp. 808-836, 2020.

[5] A. Belhadi, S. Kamble, C. J. C. Jabbour, A. Gunasekaran, N. O. Ndubisi and M. Venkatesh, 'Manufacturing and service supply chain resilience to the COVID-19 outbreak: Lessons learned from the automobile and airline industries.' Technological Forecasting and Social Change, vol. 163, pp. 120447, 2021.

[6] R. Blake, 'Advanced manufacturing's moment: Making supplies for the war on COVID-19.' Forbes. 2020. [Online]. Available: https://www.fo rbes.com/sites/richblake1/2020/04/03/advancedmanufacturings-mome nt-making-supplies-for-the-war-on-covid19/#1e6037642d1b

[7] M. Cai and J. Luo, 'Influence of COVID-19 on Manufacturing Industry and Corresponding Countermeasures from Supply Chain Perspective.' Journal of Shanghai Jiaotong University (Science), vol. 25, no. 4, pp. 409-416, 2020.

[8] F. T. Chan, N. Kumar, M. K. Tiwari, H. C. Lau and K. Choy, 'Global supplier selection: a fuzzy-AHP approach. International Journal of Production Research, vol. 46, no. 14, pp. 3825-3857, 2008.

[9] M. Chand and S. Avikal, 'An MCDM based approach for purchasing a car from Indian car market.' In 2015 IEEE Students Conference on Engineering and Systems (SCES) (pp. 1-4). IEEE. November 2015.

[10] H. Deng, 'Multicriteria analysis with fuzzy pairwise comparison.' International Journal of Approximate Reasoning, vol. 21, no. 3, pp. 215-231, 1999.

[11] K. Figueiredo, R. Pierott, A. W. Hammad and A. Haddad, 'Sustainable material choice for construction projects: A life cycle sustainability assessment framework based on BIM and Fuzzy-AHP.' Building and Environment, vol. 196, pp. 107805, 2021.

[12] Z. Güngör, G. Serhadlıoğlu and S. E. Kesen, 'A fuzzy AHP approach to personnel selection problem.' Applied Soft Computing, vol. 9, no. 2, pp. 641-646, 2009.

[13] J. L. Harris, P. Sunley, E. Evenhuis, R. Martin, A. Pike and R. Harris, 'The Covid-19 crisis and manufacturing: How should national and local industrial strategies respond? Local Economy, vol. 35, no. 4, pp. 403-415, 2020.

[14] M. Li, H. Wang, D. Wang, Z. Shao and S. He, 'Risk assessment of gas explosion in coal mines based on fuzzy AHP and bayesian network.' Process Safety and Environmental Protection, vol. 135, pp. 207-218, 2020.

[15] R. Subramoniam, F. Charnley, J. Patsavellas, D. Widdifield and K. Salonitis, 'Manufacturing in the time of COVID-19: An Assessment of Barriers and Enablers.' IEEE Engineering Management Review, vol. 48, no. 3, pp. 167-175, 2020.

[16] N. Saccani, C. Kowalkowski, M. Paiola and F. Adrodegari, 'Navigating disruptive crises through service-led growth: The impact of COVID-19 on Italian manufacturing firms.' Industrial Marketing Management, vol. 88, pp. 225-237, 2020.

[17] P. Sahoo and Ashwani, 'COVID-19 and Indian economy: Impact on growth, manufacturing, trade and MSME sector.' Global Business Review, vol. 21, no. 5, pp. 1159-1183, 2020.

[18] P. C. Shete, Z. N. Ansari and R. Kant 'A Pythagorean fuzzy AHP approach and its application to evaluate the enablers of sustainable supply chain innovation.' Sustainable Production and Consumption, vol. 23, pp. 77-93, 2020.

[19] A. R. Singh, P. K. Mishra, R. Jain and M. K. Khurana, 'Robust strategies for mitigating operational and disruption risks: a fuzzy AHP approach.' International Journal of Multicriteria Decision Making, vol. 2, no. 1, pp. 1-28, 2012.

[20] F. Torfi, R. Z. Farahani and S. Rezapour, 'Fuzzy AHP to determine the relative weights of evaluation criteria and Fuzzy TOPSIS to rank the alternatives.' Applied Soft Computing, vol. 10, no. 2, pp. 520-528, 2010.

[21] L. A. Zadeh, 'Fuzzy sets.' In Fuzzy sets, fuzzy logic, and fuzzy systems: selected papers by Lotfi A Zadeh (pp. 394-432), 1996.

[22] E. K. Zavadskas, Z. Turskis, Ž.Stević and A. Mardani, Modelling procedure for the selection of steel pipes supplier by applying fuzzy AHP method.' Operational Research in Engineering Sciences: Theory and Applications, vol. 3, no. 2, pp. 39-53, 2020.

[23] D. Ivanov and A. Dolgui, 'A digital supply chain twin for managing the disruption risks and resilience in the era of Industry 4.0.' Production Planning & Control, vol. 32, no. 9, pp. 775-788, 2021.

12

Genetic Algorithms for Selection of Critical Cytological Features in Cancer Datasets

Shona Afonso and Anusha Pai

Department of Computer Engineering, Padre Conceicao College of Engineering, India

Abstract

Feature selection using soft computing techniques is ideally used to identify and eliminate irrelevant and redundant features that do not enhance the accuracy of the prediction model. However, the process of feature selection is a daunting task, mainly due to the large search space. The domain of bionics is being explored, and metaheuristic algorithms inspired by biological processes are used to create better-designed soft computing models. This chapter compares the use of genetic algorithms combined with logistic regression and the random forest approach to optimize classification by searching for an optimal feature weightage. Selected features are used to create a neural network model and the performance metrics are assessed. It has been observed that the model constructed with the selected critical parameters exhibits a performance that is highly satisfactory and would effectively aid in the diagnosis of the disease.

Keywords: Data analysis, visualization, genetic algorithms, random forest, logistic regression, neural network.

12.1 Introduction

Various soft computing systems [1, 2] have been developed and used by healthcare professionals to diagnose and make medical decisions.

High-dimensional data analysis poses a huge challenge for researchers in the field of data mining and machine learning. The increase in the dimensionality of a domain is proportional to an increase in the count of features. All the variables in a dataset are always not useful in the construction of an efficient model. The generalization competence of the model and the overall accuracy of a classifier are reduced with redundant features. Finding an optimal set of features is intractable and problems related to its selection have been proved to be NP-hard. The method of feature selection offers an effective way to overcome this challenge by eliminating redundant and irrelevant data. The benefits derived include improvement in learning accuracy, reduction of computation time, and facilitates an enhanced understanding of the learning model.

A standard feature selection approach [3] entails the generation of a subset of features, evaluation of the subset, termination criterion, and valida-tion. Subset generation produces candidate feature subsets that are evaluated and compared with the previous best subset according to certain evaluation criteria to retain the better option. The ranking of features determines the importance of any individual feature based on statistics or some functions of the classifier's outputs.

The research work documented in the chapter attempts to identify the most critical features using the evolutionary approach of genetic algorithms. Genetic algorithms have proven to be computationally effective in situations of exhaustive search space and high-dimensional data. Further neural network models are constructed using the chosen parameters and the performance metrics analyzed.

12.2 Literature Survey

The authors in [4] have used nested genetic algorithms on microarray datasets to recognize a few genes that were significant in identifying cancer of the colon. The nested GA approach significantly improved the accuracy of clas-sification as compared to other feature selection methods such as KNN and RF. Gene selection programming based on gene expression programming has been proposed and implemented by the authors in [5] for choosing the subset of relevant genes for the efficient classification of cancer cells. The chapter reported better accuracy of classification with reduced processing time. The use of an enhanced gray wolf optimizer in combination with a support vector machine has been used in [6] to identify the subset of features necessary for distinguishing benign cancer cells from malignant ones with enhanced accuracy of classification as compared to the methods that are currently used for classification. The authors in [7] have used genetic algorithms and artificial neural networks for selecting feature subsets with improved classification performance and reduced search times. The use of horizontal scaling with particle swarm optimization has been explored in [8] for feature selection in cancer prognosis, yielding reasonable accuracy with reduced runtime. A novel technique of hybrid multi-population adaptive genetic algorithm for feature selection by overlooking the irrelevant features in high-dimensional datasets for accurate classification of tumors of various types has been explored in [9]. Feature selection problem has been dealt with using a combination of community detection approach and genetic algorithms in [10] with higher efficiency and faster convergence. The authors in [11] have investigated the use of the sine cosine algorithm and genetic algorithm for feature selection by mediating the search space exploitation and exploration strategies. By doing so, they were able to achieve maximal classification accuracy with minimal size of features on all the testing data. The use of genetic algorithms for feature selection based on histological images for pattern recognition of lymphomas was carried out in [12]. Feature selection based on a multi-stage approach for selecting CpG sites from different cancer datasets has been proposed in [13]. The approach combined three filter feature selection methods and applied genetic algorithms as a wrapper feature selection technique, resulting in improved accuracies in the classification of cancer cells.

Genetic programming and ensemble algorithms have been used to build an automated system that differentiates malignant tumors from benign ones [4]. The authors have used genetic programming in solving the issue of hyperparameters and have used accuracy and ROC curves in testing the performance of the system. Neural computing and soft computing methods for selecting the relevant features and refining the forecasting process by the use of ensemble learning for lung cancer prediction have been discussed in [15]. The developed system had a minimum error rate and was also successful in minimizing the overfitting of the cancer features. An automated system to compute cancer cells of the brain using image processing and techniques of soft computing has been investigated in [16]. The study established the superiority of the fuzzy C-mean approach to cancer cell detection with higher accuracy and lesser time complexity. Clustering algorithms have been applied to gene expression data for cancer tissue classification problems in [17] and [18]. The notion of multi-objective optimization has been applied and results indicate better performance of the classifier as compared to other single objective clustering techniques used in literature.

The research work presented in the chapter aims to create an efficient neural network using the critical features selected by the genetic algorithm with the help of different evaluators. The attempt is to create a more efficient and stable model that provides satisfactory results in correctly classifying benign and malignant cancer cases.

12.3 Research Methodology

The proposed prediction model in Figure 12.1 includes the following stages:

   i exploratory analysis and data visualization;
   ii feature selection using genetic algorithms;
  iii model design and implementation.

12.3.1 Exploratory Analysis and Data Visualization

Detailed and elaborate exploration of the dataset was carried out to identify missing values, null values, and other inconsistencies. Data were analyzed in a visual context to identify interesting patterns and correlations between attributes.

12.3.1.1 Data Analysis

The breast cancer dataset used in the study was obtained from the UCI machine learning repository. The class label is provided in terms of a

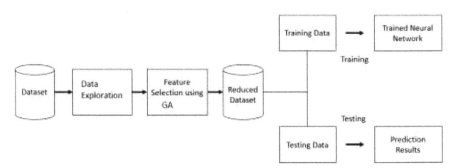

Figure 12.1   General framework for the prediction model.

B   357
M   212
Name: diagnosis, dtype: int64

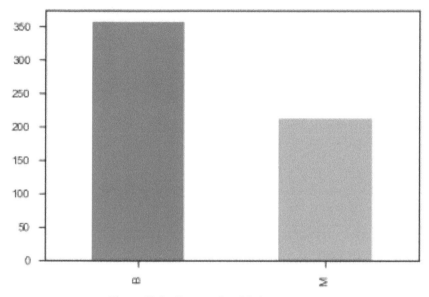

Figure 12.2   Dataset class label outcomes.

diagnostic status represented as malignant or benign. The dataset under consideration includes 30 features that detail the parameters of the cell nuclei.

Class imbalance arises when the total number of a class of data (malignant) is far less than the total number of another class of data (benign) or vice versa. Imbalance can cause cases of overfitting and underfitting, thereby resulting in models that have poor predictive performance. As shown in Figure 12.2, there are 357 benign cases highlighted in blue and 212 malignant cases specified in green, thus indicating a balanced class scenario.

12.3.1.2 Visual Analysis

A heatmap is a graphical representation of knowledge wherein the values contained in a matrix are represented as colors. A partial heatmap generated in Figure 12.3 indicates that correlation exists amongst a number of features in the dataset resulting in redundancy. It is undesirable to have highly correlated features as it causes overfitting of the model and reduces

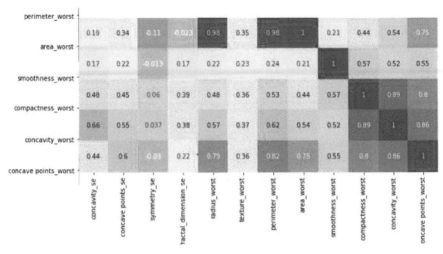

Figure 12.3   Visual analysis of the breast cancer dataset: correlation matrix.

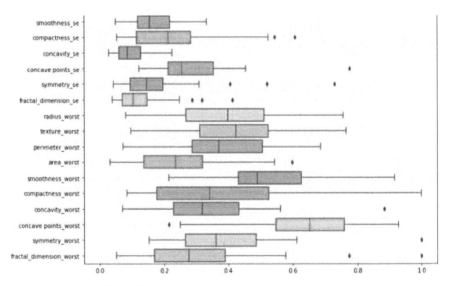

Figure 12.4   Visual analysis of the breast cancer dataset: box plot.

performance and efficiency. The positive correlation between two variables is represented by the darkness of the red color, while the negative correlation between two variables is represented by the darkness of the blue color. As indicated in Figure 12.3, perimeter worst is highly correlated to area worst

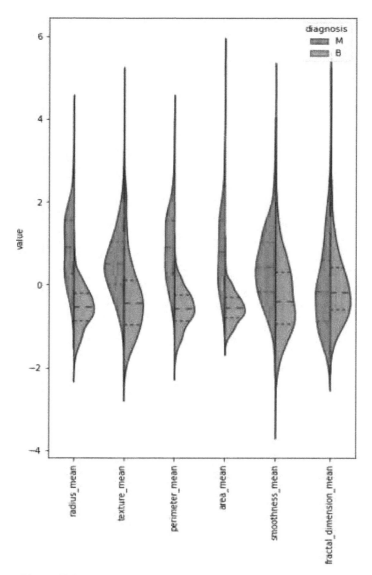

Figure 12.5 Visual analysis of the breast cancer dataset: violin plots.

and area_mean. Similarly, area worst and radius worst showed a high degree of correlation. The correlation matrix generated was very extensive, making it difficult to identify critical features. A box plot is a method for graphically demonstrating the locality, spread, and skewness of groups of numerical data through their quartiles. An analysis of the features in Figure 12.4 indicates that symmetry_se, and fractal dimension_se, have outliers.

Violin plots show the probability density of the data at different values with wider sections in violin plots representing a higher probability than the skinnier sections. This violin plot in Figure 12.5 shows the relationship of features to value.

12.3.2 Feature Selection using Genetic Algorithms

Genetic algorithms [4, 5, 9, 15] are probabilistic search algorithms that allow for the recombination of individual parameters and provide an indication of the quality of these combinations.

The following steps are carried out until the best solution is reported:

  i An initial random population is instantiated.
  ii Evaluation of a fitness function for each chromosome.
  iii Crossover operation to create new chromosomes.
  iv Mutation of genes based on the parameter value.

12.3.2.1 Comparison of Various Classification Models

Different classification models [19] were implemented on the dataset, the results of which are represented in Table 12.1. The highest accuracy is observed with classifiers logistic regression and random forest. Further assessing the other parameters, it has been observed the true positives and true negatives in both the classification models are similar. These results make logistic regression and random forest a suitable choice as evaluators to be used with genetic algorithms.

Note: TB: true benign; TM: true malignant; FM: false malignant; FB: false benign.

Table 12.1   Classification models accuracy metrics.

No.	Classifier	TB	TM	FM	FB	Accuracy
1	Logistic regression	106	59	2	4	96.491
2	Nearest neighbor	107	57	1	6	95.906
3	Kernel SVM	108	0	0	63	63.157
4	Naïve Bayes	104	57	4	6	94.152
5	Decision tree algorithm	106	58	2	5	95.906
6	Random forest classifier	107	58	1	5	96.491

12.3.2.2 Realization of Feature Importance Scores

Random forest is an ensemble-based learning algorithm that is composed of a collection of decision trees. After being fit to the training dataset, the model provides a feature importance property that can be accessed to retrieve the relative importance scores for each input feature. As indicated in Figure 12.6 (a), features numbered 3, 20, 22, 23, and 27 have higher scores as compared to the other features.

Similarly, we can fit a logistic regression model on the regression dataset and retrieve the coefficient property that contains the coefficients found for each input variable. These statistics prove that some variables are much more important than others. Figure 12.6 (b) clearly shows that there are so many variables with zero importance (or near-zero due to rounding), which should ideally be eliminated. Features numbered 21, 22, 25, and 26 have a high score as compared to the other features that have close to zero and negative valued scores.

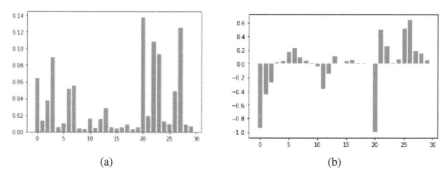

(a)   (b)

Figure 12.6   (a) Random forest classifier. (b) Logistic regression.

12.3.2.3 Comparison of random forest classifier with logistic regression as evaluators for feature selection using genetic algorithms

The idea behind the genetic-algorithm-based feature selection is to optimize the random forest and logistic regression classifiers by selecting the appropriate parameter values and improve the detection rate of the prediction model by using the optimized feature set. As shown in Figure 12.7, the highest score of 0.889 corresponds to a set of six selected features, namely "compactness_mean," "concavity_mean," "fractal_dimension_mean," "radius_worst," "concavity_worst," and "concave points_worst." Similarly, using random forest as an optimizer as indicated in Figure 12.8, it provided a set of seven selected features with a score of 0.991 corresponding to

No of Feats	Chosen Feats	Scores
6	Index(['compactness_mean', 'concavity_mean', '...	0.889
6	Index(['concavity_mean', 'concave points_mean'...	0.889
7	Index(['compactness_mean', 'concavity_mean', '...	0.889
6	Index(['concavity_mean', 'concave points_mean'...	0.889
5	Index(['compactness_mean', 'concavity_mean', '...	0.888
5	Index(['concavity_se', 'radius_worst', 'compac...	0.886
4	Index(['radius_worst', 'compactness_worst', 'c...	0.885
3	Index(['radius_worst', 'compactness_worst', 'c...	0.883
2	Index(['radius_worst', 'concavity_worst'], dty...	0.871

Figure 12.7   Logistic regression as an evaluator.

No of Feats	Chosen Feats	Scores
7	Index(['concavity_mean', 'perimeter_se', 'conc...	0.991
5	Index(['concave points_mean', 'radius_se', 'co...	0.991
5	Index(['concave points_mean', 'concavity_se', ...	0.990
4	Index(['area_se', 'area_worst', 'concavity_wor...	0.990
4	Index(['area_mean', 'concave points_mean', 'ra...	0.990
10	Index(['radius_mean', 'texture_mean', 'perimet...	0.990
4	Index(['concave points_mean', 'concave points_...	0.985
2	Index(['area_mean', 'concave points_worst'], d...	0.983
3	Index(['perimeter_mean', 'smoothness_mean', 't...	0.980

Figure 12.8   Random forest as an evaluator.

"concavity_mean," "perimeter_se," "concave points_se," "texture_worst," "area_worst," "smoothness_worst," and "symmetry_worst."

12.3.3 Neural Network Construction and Performance Analysis

The study involved constructing a feed-forward network [16], where in *m* inputs are provided to the first layer and the last layer produces a single output while learning takes place through weight adjustments and backpropagation.

12.3.3.1 Model Design Description

Two different multilayer feed-forward neural networks [7] were built, namely Model 1 with six input features and Model 2 with seven input features. A single output indicated whether the resulting class was malignant or benign. The number of neurons in the intermediate layer is 2/3 × no of input neurons. A bias value of 1 and a learning rate of 0.15 was set by default. The activation function chosen was rectified linear unit (ReLU). Binary_crossentropy was the loss function and the sigmoid function was used when dealing with classification problems with two types of results. A training set (70%), testing set (15%), and validation set (15%) were used. Hyperparameters were adjusted to achieve the best performance and accurate results [20].

12.3.3.2 Results and Observations

As observed in Figure 12.9 (a), Model 1 provided an accuracy of 96.51. The confusion matrix indicates that true benign = 60, true malignant = 23, false benign = 0, and false malignant = 3. In the case of Model 2, as shown in Figure 12.9 (b), an accuracy of 98.83 was obtained with true benign = 55, true malignant = 30, false benign = 1, and false malignant = 0.

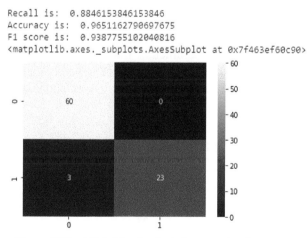

```
Recall is:    0.8846153846153846
Accuracy is:  0.9651162790697675
F1 score is:  0.9387755102040816
<matplotlib.axes._subplots.AxesSubplot at 0x7f463ef60c90>
```

Figure 12.9   (a)Model using logistic regression feature set.

```
Recall is:   1.0
Accuracy is:   0.9883720930232558
F1 score is:   0.9836065573770492
<matplotlib.axes._subplots.AxesSubplot at 0x7f463f382310>
```

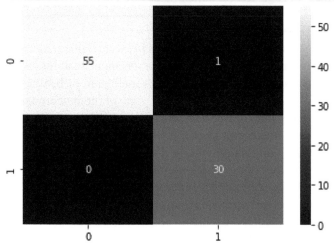

Figure 12.9   (b)Model using random forest feature set.

12.3.3.3 Exploration of Optimization of GA using Random Forest as an Estimator

A plot of the evolution of the optimization routine as shown in Figure 12.10 indicates the algorithm started with an accuracy of around 0.93 in generation 0, which generated hyperparameters randomly. But the accuracy improved while the algorithm chooses a new set of hyperparameters using evolutionary strategies. Table 12.2 shows the best values obtained for each parameter.

Table 12.2   Parameter values.

No.	Parameter	Value
1	n_estimators	117
3	max_depth	6
4	min_weight_fraction_leaf'	0.010783322895492185
5	max-leaf nodes	23
6	Bootstrap	True

Figure 12.10 Optimization routine.

12.4 Conclusion

Soft computing approaches can adapt to the problem domain making these approaches powerful, reliable, and efficient in the medical diagnostic process.

Feature selection is the key influence factor for building accurate predictive models. Therefore, it is beneficial to discard the conflicting and unnecessary features from the dataset under consideration. Developing an efficient model for disease classification was extensive and challenging since it was necessary to have critical features for model design amongst the large set of attributes available. A comparison among different classifiers enabled the selection of the most efficient classifiers to be used as evaluators with the genetic algorithm. Results revealed that the model built using features selected using random forest as an evaluator with genetic algorithms provided a higher accuracy. Further assessment of the parameters showed the best attribute values selected over multiple generations of the genetic algorithm execution. The method of genetic algorithms can therefore be adopted when standard feature selection approaches do not provide satisfactory results or the amount of data to be analyzed is enormous.

Acknowledgement

The authors express their gratitude to the reviewers for their critical input in improving the standard of the chapter.

References

[1] B. Chitra, S. S. Kumar, "Recent advancement in cervical cancer diagnosis for automated screening: a detailed review." Journal of Ambient Intelligence and Humanized Computing, 13, pages251–269, 2022.

[2] R. K. Purwar and V. Srivastava, "Recent advancements in detection of cancer using various soft computing techniques for MR images." In Progress in Advanced Computing and Intelligent Engineering (pp. 99-108). Springer, Singapore, 2018.

[3] S. A. Medjahed, T. A. Saadi, A. Benyettou and M. Ouali, "Kernel-based learning and feature selection analysis for cancer diagnosis." Applied Soft Computing, 51, 39-48, 2017.

[4] S. Sayed, M. Nassef, A. Badr and I. Farag, "A Nested Genetic Algorithm for Feature Selection in High-dimensional Cancer Microarray Dataset." Expert Systems With Applications (2018), doi:https://doi.org/10.1016/j.eswa.2018.12.022

[5] A. Russul, H. Jingyu, H. Azzawi and X. Yong, "A novel gene selection algorithm for cancer classification using microarray datasets." BMC medical genomics, 12(10), 2019.

[6] S. Kumar and M. Singh, "Breast Cancer Detection Based on Feature Selection Using Enhanced Grey Wolf Optimizer and Support Vector Machine Algorithms." Vietnam Journal of Computer Science Vol. 8, No. 2 (2021) 177–197.

[7] T. A. Mohammed , S. Alhayali , O. Bayat and O. N. Uçan, "Feature Reduction Based on Hybrid Efficient Weighted Gene

GeneticAlgorithms with Artificial Neural Network for Machine Learning Problems in the Big Data." Hindawi Scientific Programming Volume 2018, Article ID 2691759, 10 pages https://doi.org/10.1155/2018/269 1759.

[8] K. Tadist , F. Mrabti, N. S. Nikolov, A. Zahi and S. Najah, "SDPSO: Spark Distributed PSObased approach for feature selection and cancer disease prognosis." J Big Data (2021) 8:19https://doi.org/10.1186/s405 37-021-00409-x.

[9] A. K. Shukla, "Multi-population adaptive genetic algorithm for selection of microarray biomarkers." Neural Computing and Applications https://doi.org/10.1007/s00521-019-04671-2

[10] M. Rostami, K. Berahmand and S. Forouzandeh, "A novel community detection based genetic algorithm for feature selection." J Big Data (2021) 8:2 https://doi.org/10.1186/s40537-020-00398-3

[11] L. Abualigah and A. J. Dulaimi, "A novel feature selection method for data mining tasks using hybrid Sine Cosine Algorithm and Genetic Algorithm." Cluster Computing https://doi.org/10.1007/s10586-021-0 3254-y

[12] D. F. Taino, M. G. Ribeiro, G. F. Roberto, G. F. D. Zafalon, M. Z. do Nascimento, T. A. A. Tosta, A. S. Martins and L. A. Neves, "Analysis of cancer in histological images: employing an approach based on genetic algorithm." Pattern Analysis and Applications https://doi.org/10.1007/s10044-020-00931-3

[13] A. Alkuhlani, M. Nassef and I. Farag, "Multistage feature selection approach for high-dimensional cancer data." Soft Comput DOI 10.1007/s00500-016-2439-9

[14] H. Dhahri, Al. E. Maghayreh, A. Mahmood, W. Elkilani and F. Nagi, M. (2019). "Automated breast cancer diagnosis based on machine learning algorithms." Journal of healthcare engineering, 2019.

[15] P. M. Shakeel, A. Tolba, Z. Al-Makhadmeh and M. M. Jaber, 'Automatic detection of lung cancer from biomedical data set using discrete AdaBoost optimized ensemble learning generalized neural networks.' Neural Computing and Applications, 32(3), 777-790, 2020.

[16] C. Thammasakorn, C. So-In, W. Punjaruk, U. Kokaew, B. Walkham, S. Permpol and P. Aimtongkham, "Brain Cancer Cell Detection Optimization Schemes Using Image Processing and Soft Computing." In Advanced Computer and Communication Engineering Technology (pp. 171-182). Springer, Cham, 2016.

[17] S. Saha, K. Kaushik, A. K. Alok and S. Acharya, "Multi-objective semi-supervised clustering of tissue samples for cancer diagnosis." Soft Computing DOI 10.1007/s00500-015-1783-5, 2015.

[18] S. P. Maniraj and P. Sardarmaran, "Classification of dermoscopic images using soft computing techniques." Neural Computing and Applications, 33, pages13015–13026, 2021.

[19] R. Mogili, G. Narsimha and K. Srinivas, "Early Prediction of Non-communicable Diseases Using Soft Computing Methodology." In Advances in Decision Sciences, Image Processing, Security and Computer Vision (pp. 696-703). Springer, Cham.2020.

[20] S. Afonso, A. Pai and R. Sardinha, "Cancer Prediction using machine learning." https://doi.org/10.1007/978-981-16-0171-2, ISSN 2194-5357.

13

Role of AI in Various Industrial Managerial Disciplines

Arpit Malik[1], Sunitha Ratnakaram[1], Venkamaraju Chakravaram[1], and Hari Krishna Bhagavatham[2]

[1]Jindal Global Business School, O.P. Jindal Global University, India
[2]OUCCBM, Osmania University, India

Abstract

Over the last two decades, continual global demand has put constant pressure on businesses to improve with time and outperform each other. Technological advancements such as artificial intelligence, big data analytics, Internet of Things (IoT), and machine learning can be seen in every sphere of business. These advancements have dramatically improved the finesse, intricacy, and sophistication of operations. Artificial intelligence is significant because it improves human capabilities like thinking, reasoning, planning, communication, and perception by allowing software to perform these functions more effectively and efficiently. The second research paper aims to study the future aspects of artificial intelligence and how it might make a difference. However, the paper's core research focuses on the role of artificial intelligence in various business operations.

Keywords: Artificial intelligence, automation, robotics, machine learning, Industrial 4.0, robots, Internet of Things (IoT), robotics, marketing, operations and supply chain management, technological intelligence

13.1 Introduction

Businesses are struggling to cope with the large amount of data that is accessible today. Traditional business tactics can no longer be relied upon, owing to the constantly changing tastes and complexities of consumers. This continuous shifting, the rising competitiveness, and the dynamic nature of business necessitate the need for innovations. Artificial intelligence, big data analytics, the Internet of Things, and machine learning are all contributing to the rising demand and redefining the way businesses function. Not only can artificial intelligence execute tasks that human beings are incapable of executing, but it can also complement human tasks and generate outputs in a way comparable to human thinking. These advancements are contributing to the economy via contribution to productivity, may it be marketing sector, manufacturing, or supply chain management. The incorporation of artificial intelligence in the marketing sector has evolved traditional marketing into non-conventional marketing. These technologies help marketers to crunch a large amount of data from the web, social media, and emails at a faster rate, which, in turn, increases return on investment (ROI), increases personalization, helps in smarter and faster decision-making, etc. Artificial intelligence is boosting productivity, product quality, supply, and delivery processes in the manufacturing industry [10]. This aids in overall cost reduction, downtime reduction, real-time data exchange, mass customization, etc. [11]. In addition to these disciplines, there are various disciplines where artificial intelligence is making its mark, may it be HR, finance, or logistics. The study will also explore several fields where AI might influence the future, such as the health sector, military, hiring process, and law, among others.

13.2 Literature Review

13.2.1 What is Artificial Intelligence?

"Artificial intelligence is the science and engineering of making intelligent machines, especially intelligent computer programs." – John McCarthy, father of AI.

Artificial intelligence refers to machines that perform tasks by imitating human intelligence and may improve themselves over time depending on the data they collect. Artificial intelligence, unlike human intelligence, is intelligence shown by machines. It provides machines' decision-making ability to learn a task without the need for pre-programmed code. Sometimes, people might find repeated work dull at times. However, with the help of machine learning (a subpart of AI), people do not have to do a tedious job of repetitive tasks. Artificially intelligent systems (robots) are constantly doing monotonous tasks for people. Artificial intelligence's goal is to solve complex problems and produce outcomes in a way similar to how humans think. Not only that artificial intelligence can perform tasks that humans are unable of performing such as autonomously anticipating expected and unexpected events like detecting tools wearing out, detecting damaged products, etc. Artificial intelligence not only automatically detects these problems but also provides solutions without human intervention.

Business demand is always shifting, with the rising competitiveness and dynamic nature of business, necessitating the need for innovations. The demand for more complex processing solutions is getting more important as the volume of data is growing every day. As a result, many companies are turning to artificial intelligence computing techniques like deep learning, the Internet of Things, big data analytics, machine learning, and natural language processing, which are contributing to the growing need and transforming how the business operates. As a result, "technological intelligence" is defined as the ability to recognize and adapt to technological breakthroughs. In modern organizational setups, artificial intelligence has proven critical in achieving appealing operational reforms [7]. Artificial intelligence can transform the work and workplace. Today, AI accounts for around 55% of all businesses across all industries. Artificial intelligence's role in business operations is to enhance customer experience, increase revenue, increase productivity and efficiency, and move a firm ahead [9].

13.2.1.1 Artificial Intelligence in Manufacturing

Global demand is growing, placing pressure on global production and requiring manufacturers to enhance productivity and product quality [10]. In addition to that, the manufacturers face regular unexpected challenges such as machine failure, demand forecasting, inventory management, delayed product delivery, etc. The manufacturing industry has always been available to adopt newer technologies. Now with the implementation of artificial intelligence, manufacturers are beginning to employ artificial intelligence to monitor machines to gather data, learn new things, and then utilize this information in predicting machine failure, detecting anomalies, detecting product defects, etc. According to a recent survey conducted by MIT, it was revealed that 60% of manufacturers are implementing artificial intelligence and machine learning processes in their daily operations. In another similar survey conducted by Capgemini, more than half of the European manufacturers use artificial intelligence processes in manufacturing while in Japan 30% and the USA 28% [13]. According to the research conducted by Capgemini,

27% of manufacturers use artificial intelligence capabilities for maintaining quality and 29% for maintenance purpose [13].

Artificial intelligence is regarded as one of the most rapidly emerging technologies and has become an integral part of the manufacturing industry. Earlier, the interaction between robots and humans in the manufacturing environment was limited. Now, humans and robots work in close proximities in the manufacturing sector such as logistics robots, and industrial robots to help automate unergonomic tasks such as helping people move heavy parts, machine feeding, and assembly operations [16]. Artificial intelligence improves or enhances the role of robots as well as provides additional benefits like quality assurance, increased safety, process optimization, cheaper operational, and maintenance costs, helps in mass production, and so on. Artificial intelligence enhances the finesse, complexity, and sophistication of robotic tasks, resulting in a decrease in workplace accidents and reducing the routine work of humans. Robots are now frequently completing tasks that were formerly assigned to humans, owing to complexity and labor limitations [2]. Manufacturing maintenance has shifted from reactive to preventative maintenance, thanks to artificial intelligence enabled predictive capabilities. Manufacturing units are employing enhanced monitoring and auto-correction of processes to improve their manufacturing processes, which aided in getting better yields and identification of inefficient machines, which contributes to the reduction in manufacturing cost [12]. For example, to improve product quality, advanced artificial intelligence technology like computer vision is used to investigate flaws in manufactured products.

Similarly, the industrial Internet of Things helps in the expansion and streamlining of industrial processes such as manufacturing, monitoring, and supply chain management due to which there is an increase in efficiency, scalability, and accessibility [15]. The industrial Internet of Things improves existing production and supply chain monitoring systems, which help in reducing overall cost, reducing downtime, and improving quality, and help to share real-time data, improving mass customization, etc. All the above-mentioned challenges and advantages reflect upon the importance of the role of artificial intelligence in manufacturing [11]. Now, managers and workers can focus on driving innovation and transforming their business.

13.2.1.2 Artificial Intelligence in Marketing

"Marketing is the activity, set of institutions, and processes for creating, communicating, delivering, and exchanging offerings that have value for customers, clients, partners, and society at large" [4].

In today's marketing, artificial intelligence plays a crucial role. Artificial intelligence has emerged as a key component of digital marketing strategies, and it is an effective tool. AI marketing is a technique for gathering data, gaining consumer insights, predicting future customer behavior, and making automated marketing decisions. Artificial intelligence is assisting marketing and companies in providing a wonderful consumer experience. Most well-known companies and services, such as Amazon, Facebook, Netflix, and Google, are constantly powered by AI algorithms. Artificial intelligence has several advantages and the potential to solve many of today's issues. Artificial intelligence helps companies to remain competitive in today's marketing environment [14]. Some of the advantages of artificial intelligence are as follows.

1. Artificial intelligence is transforming the way of marketing by helping marketers to crunch a huge amount of data from the web, social media, and emails at a faster rate.
2. Increases return on investment: Artificial intelligence also provides digital solutions for acquiring and retaining customers. Artificial intelligence helps organizations to analyze real-time customer data, allowing them to instantly respond to customer needs, which, in turn, increases return on investment.
3. Minimizes human error: Artificial intelligence is used for limiting human intervention, which, in turn, helps in minimizing human errors.
4. Increased personalization: Artificial intelligence will make marketing more personalized; for instance, an e-commerce shop receives more favorable reactions by sending push notifications or making product recommendations to the customers. An increase in personalization will help marketers to attract more customers as it will make customers feel more special and they tend to get influenced due to it.
5. Smarter and faster decision-making: Artificial intelligence provides real-time data analysis, which helps marketers to reach a decision or modify according to the changes observed before the campaign ends.
6. There is a fierce rivalry and technological disruptions that are changing business operations. A customer-centric strategy is critical for an organization's success. According to a survey, 76% of customer wants or expects businesses to understand their needs and expectations. In 2011, there were 150 marketing tools accessible on the market; by 2017, that number had risen to 5000, and now, the market has even more tools. Small- and medium-sized enterprises may now employ AI, thanks to

the accessibility and cost of marketing solutions. Artificial intelligence now accounts for about 55% of all businesses across all industries. This is because marketers are increasing their return on investment in a short period by acquiring customer insights, and today's customers also want businesses to understand and fulfill their requirements. As a result, businesses are always pushing to shift their operations to artificial intelligence.

13.2.1.3 Significance of Artificial Intelligence in Supply Chain Management

According to a recent study, the implementation of artificial intelligence in supply chain and operations is generating the most revenue for businesses. Supply chain management involves managing the overall flow of raw materials and guarantees that customers get a high-quality finished product on schedule.

- Artificial intelligence enables increased contextual intelligence: Artificial intelligence enabled technologies such as intelligent robotic sorting and artificial intelligence powered visual inspection to provide the information needed to reduce operating costs and inventories while also responding to clients more swiftly [6].
- Enhances productivity: Artificial intelligence combines the capacity of complex technologies like unsupervised learning and supervised learning to discover key aspects and challenges affecting supply chain performance [6].
- Enhancing accuracy of demand forecasting: Previous technologies failed to account for a variety of demand-side factors such as consumer preferences, resulting in unreliable results. Now, artificial intelligence provides forecasts based on real-time sales, weather, and other factors by which warehouse management can be enhanced and demand forecasting accuracy can be improved. Amazon facilities, for example, now deploy autonomous ground vehicles, self-driving forklifts, and automated sorting systems to cut costs [6].
- Enhanced customer experience: Artificial intelligence personalizes the customer experience; for example, when someone orders a package from amazon, they offer voice-based service to track down their packages and information is provided through Echo (powered by Amazon Alexa) [6].
- Improving production planning and scheduling: Before artificial intelligence, production planning and scheduling was a difficult and hectic

task. But now with artificial intelligence, we can easily analyze and optimize a wider range of parameters [6].

13.2.1.4 Other Disciplines Where Artificial Intelligence Is or Can Play a Crucial Role

1. HR department: From employee onboarding through performance review, HR is in charge. The majority of a manager's time is spent on administrative, coordinating, and control responsibilities. In HR, AI is used to automate operations like meeting scheduling, performance appraisal, salary processing, and replying to employee inquiries, among others. Furthermore, artificial intelligence may be used during job interviews to respond to applicant inquiries, shortlist resumes, and monitor candidate facial expressions during the interview [1].

2. Finance department: Artificial intelligence can be used to optimize the efficiency of the finance department. Artificial intelligence can be used in fraud detection, risk management, managing customer data, and automated virtual financial assistants.

   - Fraud detection: Artificial intelligence can be used to analyze financial transactions and help analysts to determine if they are fraudulent or not, which will help in identifying rarity, the warning signs of fraud attempts, and incidences [1].
   - Risk management: Artificial intelligence has the potential to make the loan approval process more efficient. It can assist in identifying the possible risk associated with providing the loan. Furthermore, artificial intelligence can assist in the analysis of data about recent transactions, market patterns, and financial activity [1].
   - Managing customer data: Technologies like data mining, text analytics, and natural language processing (NLP) can help in retrieving information accurately and efficiently [1].
   - Automated virtual financial assistant: Robo advisors are systems that are commonly used nowadays by fintech start-ups and financial companies. They provide recommendations regarding buying and selling of stocks and bonds [1].

3. Logistics: The unorganized and fragmented logistics business has been altered by artificial intelligence, blockchain, and big data. Artificial intelligence helps save time and money, improves accuracy, and increases productivity. Predictive analytics, driverless cars, and smart highways are all examples of how these technologies have changed the logistics

industry. It aids in the automation of several time-consuming operations as well as demand forecasting, saving both time and money. Artificial intelligence aids in logistics route optimization, resulting in increased profitability and lower shipping costs. By gathering and analyzing data to make informed decisions, artificial intelligence is transforming warehousing operations. Furthermore, robots are altering the way inventory is located, tracked, and moved in warehouses, saving time [8].

13.3 S.W.O.T. Analysis:

13.4 Managerial Implications

Administrative work contributes to a major portion of every manager's daily responsibility. Artificial intelligence may automate mundane management tasks, allowing manager's focus on high-value tasks; for example, artificial-intelligence-enabled systems can notify management about any potential danger or bottlenecks in the process. Workers feel robots are better at

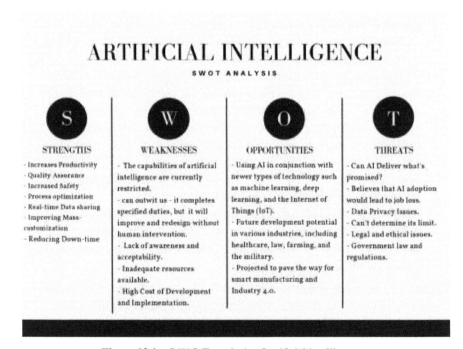

Figure 13.1   S.W.O.T. analysis of artificial intelligence.

delivering fair information, managing work schedules, problem-solving, and budget management than their managers, according to the poll, but managers are better at empathy, mentoring, and building a work culture. Artificial intelligence may help managers finish repetitive tasks, enhance customer relationships by delivering rapid replies (since AI can efficiently analyze massive data sets), and improve team dynamics by offering immediate feedback to team members. There is a disconnection between the willingness and actual adoption of artificial intelligence. To overcome this gap, artificial intelligence assists managers in moving away from traditional marketing methods and toward digital marketing. Artificial intelligence can be used to help managers segment audiences more accurately. Knowing segments accurately will assist in personalizing their content. Acting through real-time AI-delivered insights is another way to bridge this gap. IBM Watson, Albert, for instance, helps with everyday planning, executing, and reporting of campaigns. It helps with everything from optimization to execution. Hilton is an early adopter of artificial intelligence. They are using it in streamlining their recruiting process; in addition to that, they are now expanding its scope to other areas. Digital assistants are used to provide feedback and performance evaluation. Foresighted managers are excited about artificial intelligence's potential applications; they see artificial intelligence as a tool that can help them advance their careers. Artificial intelligence has the potential to assist excellent managers in becoming exceptional.

13.5 Discussion

The future of artificial intelligence in industries is yet unknown. The technology is still in its development stage, and there is much to learn and improve. Currently, factories automate processes and machinery through a fixed approach, meaning robots cannot make or select an approach to make decisions. In the future, AI will help manufacturers to develop approaches to handle unforeseen scenarios. In the future, manufacturers will be able to prepare defect-free production processes. When artificial intelligence solutions are combined with additional supporting technologies such as machine learning, industrial Internet of Things, devices, and platforms, smart manufacturing and Industry 4.0 will become a reality [10]. Artificial intelligence can help in providing more scope for customization and real-time analysis, reducing the cost of single-run and small-batch goods.

As we discussed some of the examples above, we can say that artificial intelligence definitely has, in some way, reduced extra efforts and has reduced the extra time for businesses in their day-to-day activities. But there are still some domains where artificial intelligence can make a significant impact.

1. Health sector: In health organizations, artificial intelligence may analyze large data sets like patient records, results of medical examinations, or clinical study data to identify hidden patterns and insights that humans would find difficult to identify on their own. In addition to that, artificial intelligence has the potential to revolutionize the healthcare industry through automated image diagnosis, robot-assisted surgery, preliminary diagnosis, administrative workflow assistance, virtual nursing assistants, and so on [5].

2. Military sector: The implementation of artificial intelligence in the military will provide commanders with significant strategic insights and will also help governments reduce combat losses by providing highly accurate real-time predictive inputs.

3. Hiring process: The recruiting process can be subjective; factors such as skin color, gender identity, and personal attractiveness standards may influence the hiring authority's choice. These characteristics might provide applicants with an unfair advantage or disadvantage. An impartial artificial intelligence algorithm that is confined to the job requirements may be constructed, resulting in a substantial change, and ensuring that the results are not influenced by subjective variables and that only the most qualified candidates are chosen for the position.

4. Law: AI may be used to understand court proceedings and will aid courts in the translation of court papers into multiple languages, as well as supporting the judiciary in case of scheduling, therefore reducing their burden. AIs can also provide useful information regarding undertrials to decision-makers when making judgments like bail, parole, and punishment.

5. Farming sector: Including artificial intelligence in farming will add to the productivity level, simplification of the process, faster output, etc. Some of the key benefits of using AI in farming are as follows:

   - Artificial intelligence can be used to identify nutrient deficiencies in the soil.
   - Farmers will be able to use artificial intelligence customized plans for their lands. Artificial intelligence will help the farmer to be connected to the world market and maintain sustainable growing production.

- Artificial intelligence can use satellite images for the identification of pests. It uses the comparisons of historical data and algorithms. This will send an alert to the farmers via smartphones for which required precautions can be taken.

13.6 Conclusion

This sudden and rapid expansion of digitization and transformation is putting pressure on companies, whether it is manufacturing, energy, or transportation, to use AI and machine learning to assist enhance operational efficiencies [17]. Artificial intelligence is the future of business, and companies should embrace AI as a guiding light if they do not want to miss out on the opportunity [3]. Nowadays, manufacturers are pursuing a mix of artificial intelligence, machine learning, and the industrial Internet of Things to increase productivity and predictive maintenance, reducing mundane and repetitive tasks to reduce downtime and improve employee satisfaction.

These technologies can also assist in creating opportunities for various degrees of optimization in logistics, warehousing, and last-mile delivery [17]. This is barely scraping the surface of AI's potential. There is more to come – research is going on to broaden the scope of artificial intelligence. Various sectors such as law, healthcare, farming, and the military are looking into its possibilities. While there are several obstacles to overcome, artificial intelligence has the potential to solve many of today's issues. But this transformation can only occur if individuals and companies collaborate to build a man–machine hybrid that is more powerful than any entity working alone. We cannot predict what the future will be, but one thing is for sure – artificial intelligence will play a significant role in shaping it. We shall soon be seeing the emergence of the "augmented age." Companies must use artificial intelligence to remain competitive in today's industrial environment [14]. Technological advancements are happening, but it is up to us to implement artificial intelligence technology if we plan on running a successful business.

Acknowledgement

I would like to thank my mentor Professor Sunitha Ratnakaram for providing me with this opportunity to work on this Project. Her expertise benefited me greatly in gaining knowledge. I have always found AI intriguing, but the project has broadened my scope. The project helped me gain a broader view of the use of artificial intelligence technology and its future in different sectors.

Lastly, I would want to express my gratitude to my parents for their assistance and unwavering support; without them, this project would not have been possible. Finally, I would want to convey my appreciation to my friends for always believing in me and supporting me.

References

[1] V. S. Mohan, (16 March, 2020) 'The emerging role of AI in business management optimization.' Retrieved from accubits vlog: https://blog.accubits.com/role-of-ai-in-business-management/ #:~:text=%20Possibilities%20of%20AI%20in%20business %20management%20optimization,operational%20efficiency%20of %20Finance%20and%20Ops...%20More%20.

[2] Wikipedia contributors. (2021, October 16). cobot. Retrieved from Wikipedia, The Free Encyclopedia.: https://en.wikipedia.org/w/ind ex.php?title=Cobot&oldid=1050185187.

[3] AI. (2020, July 16). What is Artificial Intelligence (AI) in Business? Retrieved from Business world it: https://www.bu sinessworldit.com/ai/artificial-intelligence-in-business /#:~:text=Artificial%20intelligence%20in%20business %20sim- ply%20involves%20the%20use,growth%20and%20transformation. %20Why%20should%20companies%20use%20AI%3F.

[4] American Marketing Association. (2013). American Marketing Associ- ation. Retrieved from www.ama.org: https://www.ama.org/the-definitio n-of-marketing-what-is-marketing/.

[5] C. Arsene, (2021, July 8). artificial intelligence in healthcare. Retrieved from health care weekly: https://healthcareweekly.com/artificial-intell igence-in-healthcare/.

[6] L. Benton, (2018, September 27). 6 WAYS AI IS IMPACTING THE SUPPLY CHAIN. Retrieved from supply chain beyond: https://supply chainbeyond.com/6-ways-ai-is-impacting-the-supply-chain/.

[7] P. Dhamija and S. Bag, 'Role of artificial intelligence in operations environment: a review and bibliometric analysis.'2019. TQM, file:///C:/Users/arpit/Downloads/2020_TQM_RoleofArtificial Intelli- genceinOperationsEnvironmentAReviewandBibliometricAnalysis.pdf.

[8] U. Gupta, (2021, June 02). Role and Importance of AI in Supply Chain Management. Retrieved from Indian retailer: https://www.indianretailer

.com/article/technology/digital-trends/role-and-importance-of-ai-in-su
pply-chain-management.a7073/.

[9] ICSID. (2021, November 14). Does Ai Help Your Company How Artificial Intelligence Helps Businesses? Retrieved from ICSID: https://www.icsid.org/uncategorized/does-ai-help-your-company-how-artificial-intelligence-helps-businesses/.

[10] I. Karabejovic, E. Karabejovic, A. Kovacevic, L. B. Mehmedovic and P. Dasic, 'Smart Sensors: Support for the Implementation of Industry 4.0 in Production Processes.' In Handbook of Research on Integrating Industry 4.0 in Business and Manufacturing (pp. 31-52).2020. IGI Global Publisher. Retrieved from https://www.igi-global.com/chapter/smart-sensors/252363.

[11] R. King, (2019, Jan 24). 7 benefits of AI in manufacturing. Retrieved from rowse: https://www.rowse.co.uk/blog/post/7-manufacturing-ai-benefits

[12] M. Kumar, (2019, Feb 25). The Role of Artificial Intelligence in Manufacturing: 15 High Impact AI Use Cases. Retrieved from data science central: https://www.datasciencecentral.com/profiles/blogs/the-role-of-artificial-intelligence-in-manufacturing-15-high.

[13] Mckinsey. (2018, June 1). AI, automation, and the future of work: Ten things to solve for. Retrieved from mckinsey global institute: https://www.mckinsey.com/featured-insights/future-of-work/ai-automation-and-the-future-of-work-ten-things-to-solve-for.

[14] S. Sharma, R. Sharma, S. Deba and D. Maitra, 'Artificial Intelligence in Marketing: Systematic review and future research directions.' International Information Management Data Insights, 2021.

[15] Terralogic. (n.d.). IoT and how IIoT (Industrial Internet of Things) became so important? Retrieved from terralogic: https://www.terralogic.com/iot-and-how-iiot-became-so-important/.

[16] the manufacturer. (2015, November 12). I, Cobot: Future collaboration of man and machine. Retrieved from the manufacturer: https://www.themanufacturer.com/articles/i-cobot-future-collaboration-of-man-and-machine/.

[17] P. Vashistha, (2020, November 21). Role Of AI And Machine Learning In Logistics Industry. Retrieved from inc42: https://inc42.com/resources/role-of-ai-and-machine-learning-in-logistics-industry/#:~:text=%20Role%20Of%20AI%20And%20Machine%20Learning%20In,Learning%20aids%20in%20bringing%20truckloads%20of...%20More%20.

14

Evaluating Fuzzy System Reliability using a Time-Dependent Hexagonal Fuzzy Number

Pawan Kumar[1] and Deepak Kumar[2]

[1]Department of Mathematics, Government Degree College, Bhikiyasain (Almora) India
[2]Department of Mathematics, D. S. B. Campus, Kumaun University, Nainital India

Abstract

For many years, the fuzzy set theory has been used in system reliability analysis. Using an introduced approach, we formulate the membership function of fuzzy reliability where a time-dependent hexagonal fuzzy number is used to represent the parameter of the failure rate of every component of the systems. The main benefit of using hexagonal fuzzy numbers in reliability theory is that a hexagonal fuzzy number has six parameters, which gives better results than triangular and trapezoidal fuzzy numbers. Thus, in this chapter, a fuzzy reliability function is proposed to formulate its membership and the uses of α-cut time-dependent hexagonal fuzzy numbers are introduced. Two numerical examples are given to exemplify the methodology for fuzzy reliability evaluation of systems.

Keywords: α-cut set, failure rate, Fuzzy reliability, time-dependent hexagonal fuzzy number.

14.1 Introduction

The most important engineering task for the design and development of any engineering system is reliability. The reliability of any system may be determined through testing or access to functional data. The reliability evaluation could be very difficult with good purity in many situations. In the appraisal of the reliability of many new systems, the component reliabilities are not known appropriately. Reliability evaluation of conceivable systems with inadequate and unspecific data on failure rate is our major anxiety. In many real-life situations, the calculation of exact values of reliabilities is very arduous in many systems due to the incertitude and inadequacy of data. Fuzzy set theory is used for dealing with this type of imprecise and inadequate data. To tackle this situation, in order to measure the fuzzy reliability of systems, fuzzy set theory is used [1]. Chen [10] explained a new way to analyze system reliability using fuzzy numbers arithmetic operations. The fuzzy reliability theory was introduced by several authors [3–5]. An assessment of fuzzy reliability using intuitionistic trapezoidal fuzzy numbers has been defined by [14] and applied to new arithmetic operations of intuitionistic fuzzy numbers. They are considered in the form of time-dependent triangular fuzzy numbers. Aliev and Kara [6] developed a way to construct the adhesive function of fuzzy reliability. Mahapatra and Roy [13] focused on the fuzzy reliability of parameters expressed as intuitionistic fuzzy numbers with some different operators. When the failure rate of any component is in the form of a time-dependent triangular intuitionistic fuzzy number, Kumar *et al.* [7] have developed and evaluated the fuzzy reliability of different systems. Sudha *et al.* [11] gave new operations on hexagonal fuzzy numbers. Kumar and Singh [9] gave a novel approach for emphasizing fuzzy reliability of systems using Weibull lifetime distribution and triangular intuitionistic fuzzy numbers. Dhurai and Karpagam [12] created a new hexagonal fuzzy number membership function. Kumar and Kumari [8] emphasized the fuzzy reliability function of systems using exponential distribution and the lifetime parameter of exponential distribution taken as hexagonal fuzzy numbers.

The present study investigates a general process introduced to analyze the fuzzy reliability of different systems. This study introduces the concept of time-dependent hexagonal fuzzy numbers and the failure rate is treated as a time-dependent hexagonal fuzzy number, which is used to build the membership function of the fuzzy reliability. A time-dependent hexagonal fuzzy number is used as the failure rate parameter for each component of this chapter to calculate the membership function of fuzzy reliability of various systems.

14.2 Preliminaries

14.2.1 Definition ([2]).

A definition of fuzzy set Z. If the set X is fixed, then:

$$\tilde{Z} = \left\{ \langle x, \mu_{\tilde{Z}}(x) \rangle : x \in X \right\},$$

where $\mu_{\tilde{Z}}(x) \in [0,\ 1]$ is the membership grade of element $x \in X$.
The time-dependent fuzzy set $\tilde{Z}(t)$ of the set X is defined as

$$\tilde{Z}(t) = \left\{ \langle x, \mu_{\tilde{Z}(t)}(x) \rangle \right\},$$

where $\mu_{\tilde{Z}(t)}(x)$ is the dynamic membership function for $x \in X$.

14.2.2 α-Cut set of time-dependent fuzzy set

The definition of α-cut set of a time-dependent fuzzy set $\tilde{Z}(t)$ is as follows:

$$\tilde{Z}_\alpha(t) = \left\{ \langle x : \mu_{\tilde{Z}(t)}(x) \geq \alpha, x \in X \rangle \right\}, \quad \alpha \in [0, 1].$$

The time-dependent fuzzy number $\tilde{Z}(t)$ is shown in Figure 14.1, which is both convex and normal. Let us suppose that the fuzzy number at time t_1 is $\tilde{Z}(t_1)$ and at time t_2 is fuzzy number $\tilde{Z}(t_2)$.

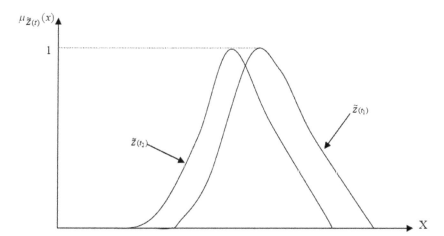

Figure 14.1   Time-dependent fuzzy number.

14.2.3 Hexagonal fuzzy numbers [11]

A fuzzy number is a hexagonal fuzzy number $\tilde{A}$ given by

$$\tilde{A} = (a_1, a_2, a_3, a_4, a_5, a_6),$$

where a_1, a_2, a_3, a_4, a_5, and a_6 are real numbers, and the degree of membership $\mu_{\tilde{A}}(x)$ of hexagonal fuzzy number $\tilde{A}$ is given as follows:

$$\mu_{\tilde{A}}(x) = \begin{cases} \frac{1}{2}\left(\frac{x-a_1}{a_2-a_1}\right), & \text{for } a_2 \geq x \geq a_1 \\ \frac{1}{2} + \frac{1}{2}\left(\frac{x-a_2}{a_3-a_2}\right), & \text{for } a_3 \geq x \geq a_2 \\ 1, & \text{for } a_4 \geq x \geq a_3 \\ 1 - \frac{1}{2}\left(\frac{x-a_4}{a_5-a_4}\right), & \text{for } a_5 \geq x \geq a_4 \\ \frac{1}{2}\left(\frac{a_6-x}{a_6-a_5}\right), & \text{for } a_6 \geq x \geq a_5 \end{cases}$$

Figure 14.2 shows hexagonal fuzzy numbers.

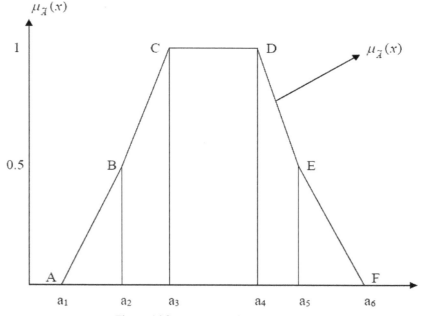

Figure 14.2 Hexagonal fuzzy numbers.

14.2.4 α-Cut hexagonal fuzzy number

$$A_\alpha = \begin{cases} [P_1(\alpha), P_2(\alpha)], & \text{for } \alpha \in [0, 0.5) \\ [Q_1(\alpha), Q_2(\alpha)], & \text{for } \alpha \in [0.5, 1] \end{cases}$$

$$A_\alpha = \begin{cases} [2\alpha(a_2 - a_1) + a_1, -2\alpha(a_6 - a_5) + a_6], & \text{for } \alpha \in [0, 0.5) \\ [2\alpha(a_3 - a_2) - a_3 + 2a_2, -2\alpha(a_5 - a_4) + 2a_5 - a_4], \\ \text{for } \alpha \in [0.5, 1]' \end{cases}$$

where $P_1(x) = \frac{1}{2}\left(\frac{x-a_1}{a_2-a_1}\right)$, $P_2(\alpha) = \frac{1}{2}\left(\frac{a_6-x}{a_6-a_5}\right)$, $Q_1(\alpha) = \frac{1}{2} + \frac{1}{2}\left(\frac{x-a_2}{a_3-a_2}\right)$, and $Q_2(\alpha) = 1 - \frac{1}{2}\left(\frac{x-a_4}{a_5-a_4}\right)$.

14.2.5 Time-dependent hexagonal fuzzy number

A time-dependent hexagonal fuzzy number $\tilde{A}(t)$ is as follows:

$$\tilde{A}(t) = (r(t) - \lambda(t) - \mu(t), r(t) - \lambda(t), r(t), s(t), s(t) + \gamma(t), s(t) + \gamma(t) + \delta(t),$$

where $p(t), q(t) \in R$ is the center, $\lambda(t) > 0$ and $\mu(t) > 0$ are the left spreads of the membership of $\tilde{A}(t)$ at time t, and $\gamma(t) > 0$ $\delta(t) > 0$ are right spreads of the membership of $\tilde{A}(t)$ at time t. Figure 14.3 shows a time-dependent hexagonal fuzzy number.

14.3 Problem Formulation

Consider two crisp sets X and Y. The failure rate parameters are in the form of a time-dependent hexagonal fuzzy set $\tilde{H}(t)$. The parameter $\tilde{H}(t)$ is given as follows:

$$\tilde{H}(t) = \{(\lambda, \mu_{\tilde{H}(t)}(\lambda)) | \lambda \in X\}. \tag{14.1}$$

The α-cut set $\tilde{H}(t)$ is calculated by the following equation:

$$\tilde{H}_\alpha(t) = \{\alpha \in X | \mu_{\tilde{H}(t)}(\lambda) \geq \alpha\}. \tag{14.2}$$

Note that $\tilde{H}_\alpha(t)$ is a crisp set.

Suppose that $\tilde{H}(t)$ is a time-dependent hexagonal fuzzy number. Then for each choice of α-cut, we have the interval

$$\tilde{H}_\alpha(t) = \begin{cases} [p_{1\alpha}(t), p_{2\alpha}(t)], & \text{for } \alpha \in [0, 0.5) \\ [q_{1\alpha}(t), q_{2\alpha}(t)], & \text{for } \alpha \in [0.5, 1] \end{cases}, \tag{14.3}$$

where $P_{1\alpha}(t)$ and $P_{2\alpha}(t)$ are increasing functions of α when $\alpha \in [0, 0.5)$and $q_{1\alpha}(t)$ and $q_{2\alpha}(t)$ are increasing functions of α when $\alpha \in [0.5, 1]$. The bounds of the interval are functions of α and $P_{1\alpha} = \min \tilde{H}_\alpha(t)$; $\alpha \in [0, 0.5)$, $q_{1\alpha} = \min \tilde{H}_\alpha(t)$; $\alpha \in [0.5, 1]$ $P_{2\alpha} = \max \tilde{H}_\alpha(t)$; $\alpha \in [0, 0.5)$, $q_{2\alpha} = \max \tilde{H}_\alpha(t)$; $\alpha \in [0.5, 1]$ respectively.

Suppose $\Psi: X \to Y$ is a continuous and bounded function from X to Y. Now we must apply to the set X to determine the fuzzy reliability that was messed up on Y. By applying Ψ to the set $\tilde{H}(t)$, we can now determine the fuzzy reliability that was messed up on Y. Now we write $y = \Psi(\lambda)$, when $y \in Y, \lambda \in X$, and $\tilde{R}(t) = \{(y, \mu_{\tilde{R}(t)}(y)) | y = \Psi(\lambda)\}$. Then the membership grade of the fuzzy reliability $\tilde{R}(t)$ using the extension principle is calculated as

$$\mu_{\tilde{R}(t)}(y) = \sup \left\{ \mu_{\tilde{H}(t)}(\lambda) : y = \psi(\lambda), \lambda \in X \right\}. \qquad (14.4)$$

If $\tilde{H}(t)$ is a normal convex set and ψ is bounded, then$\tilde{R}(t)$is also a normal and convex set; therefore, we can calculate the corresponding interval

$$\Psi(\lambda) = y \in [a_{1\alpha}(t), a_{2\alpha}(t)], \text{where} \lambda \in [P_{1\alpha}(t), P_{2\alpha}(t)], \quad \text{for } \alpha \in [0, 0.5)$$
$$(14.5a)$$

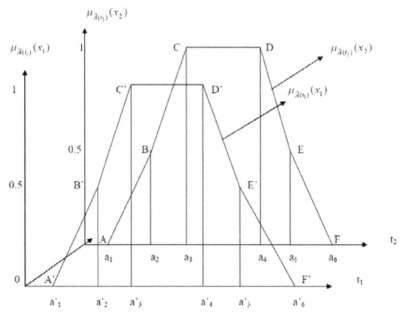

Figure 14.3   Time-dependent hexagonal fuzzy number.

and

$$\Psi(\lambda) = y \in [b_{1\alpha}(t), b_{2\alpha}(t)], \text{where} \lambda \in [q_{1\alpha}(t), q_{2\alpha}(t)], \quad \text{for } \alpha \in [0.5, 1] \tag{14.5b}$$

where $a_{1\alpha}(t)$ and $a_{2\alpha}(t)$ are min and max of ψ over $\tilde{H}(t)$ for $\alpha \in [0, 0.5)$ respectively, i.e.,

$$a_{1\alpha}(t) = \min \Psi(\lambda); p_{1\alpha}(t) \leq \lambda \leq p_{2\alpha}(t). \tag{14.6a}$$

$$a_{2\alpha}(t) = \max \Psi(\lambda); p_{1\alpha}(t) \leq \lambda \leq p_{2\alpha}(t) \tag{14.6b}$$

and $b_{1\alpha}(t)$ $b_{2\alpha}(t)$ are min and max of Ψ over $\tilde{H}(t)$ for $\alpha \in [0.5, 1]$, respectively, i.e.,

$$b_{1\alpha}(t) = \min \psi(\lambda); q_{1\alpha}(t) \leq \lambda \leq q_{2\alpha}(t) \tag{14.7a}$$

$$b_{2\alpha}(t) = \max \psi(\lambda); q_{1\alpha}(t) \leq \lambda \leq q_{2\alpha}(t). \tag{14.7b}$$

If $a_{1\alpha}(t)$ and $a_{2\alpha}(t)$ are invertible with respect to α for $\alpha \in [0,0.5)$, then the left and right shape functions are $f_{\tilde{R}(t)}(y)$ and $g_{\tilde{R}(t)}(y)$, respectively, can be acquired as

$$f_{\tilde{R}(t)}(y) = [a_{1\alpha}]^{-1} = \left[\min_{y_1 \leq y \leq y_2} y \right]^{-1}, \tag{14.8}$$

$$g_{\tilde{R}(t)}(y) = [a_{2\alpha}]^{-1} = \left[\max_{y_2 \leq y \leq y_3} y \right]^{-1}, \tag{14.9}$$

and if $b_{1\alpha}(t)$ and $b_{2\alpha}(t)$ are invertible with respect to α for $\alpha \in [0.5,1]$, then the left and right shape functions are $h_{\tilde{R}(t)}(y)$ and $k_{\tilde{R}(t)}(y)$, respectively, and can be acquired as

$$h_{\tilde{R}(t)}(y) = [b_{1\alpha}]^{-1} = \left[\min_{y_{41} \leq y \leq y_5} y \right]^{-1}, \tag{14.10}$$

$$k_{\tilde{R}(t)}(y) = [b_{2\alpha}]^{-1} = \left[\max_{y_5 \leq y \leq y_6} y \right]^{-1}, \tag{14.11}$$

and the membership function can be constructed as

$$\mu_{\tilde{R}(t)}(y) = \begin{cases} f_{\tilde{R}(t)}(y), & y_1 \le y \le y_2 \\ g_{\tilde{R}(t)}(y), & y_2 \le y \le y_0 \\ 1, & y_3 \le y \le y_4 \\ h_{\tilde{R}(t)}(y), & y_4 \le y \le y_5 \\ k_{\tilde{R}(t)}(y), & y_5 \le y \le y_6 \end{cases} , \tag{14.12}$$

where $f_{\tilde{R}(t)}(y_1) = k_{\tilde{R}(t)}(y_6) = 0$.

14.4 Reliability evaluation with time-dependent hexagonal fuzzy number

In reliability theory, the reliability function of any system in terms of failure rate can be defined as follows:

$$\tilde{R}(t) = \exp\left[-\int_0^t \tilde{\lambda}(t')dt'\right], t > 0, \tag{14.13}$$

where $\tilde{\lambda}(t')$ is the fuzzy failure rate parameter.

Let us suppose that the failure rates of any system can be considered as time-dependent hexagonal fuzzy numbers as

$$\tilde{H}(t) = (r(t) - \mu(t) - \nu(t), r(t) - \lambda(t), r(t), s(t), s(t) + \varepsilon(t), s(t) + \varepsilon(t) + \zeta(t)).$$

The α-cut set of the fuzzy failure rate $\tilde{H}(t)$ is as follows:

$$\tilde{H}_\alpha(t) = \begin{cases} [P_1(\alpha), P_2(\alpha)], & \text{for } \alpha \in [0, 0.5] \\ [Q_1(\alpha), Q_2(\alpha)], & \text{for } \alpha \in [0.5, 1] \end{cases}$$

$$= \begin{cases} [r(t) - \mu(t) - \nu(t) + 2\alpha\nu(t), s(t) + \varepsilon(t) + \zeta(t) - 2\alpha\zeta(t)], \\ \quad \text{for } \alpha \in [0, 0.5) \\ [r(t) - 2\mu(t) + 2\alpha\mu(t), s(t) + 2\varepsilon(t) - 2\alpha\varepsilon(t)], \text{for } \alpha \in [0.5, 1]. \end{cases}$$
$$\tag{14.14}$$

Now, using formula (14.2), we obtained

$$a_{1\alpha}(t) = \min\left[\exp\left(-\int_0^t \lambda(t')dt'[\right)\right]$$
$$\text{s.t. } r(t) - \mu(t) - \nu(t) + 2\alpha\nu(t) \le \lambda(t) \le s(t) + \varepsilon(t) + \zeta(t) - 2\alpha\zeta(t)$$
$$a_{2\alpha}(t) = \max\left[\exp\left(-\int_0^t \lambda(t')dt'\right)\right]$$

s.t. $r(t) - \mu(t) - \nu(t) + 2\alpha\delta(t) \leq \lambda(t) \leq s(t) + \varepsilon(t) + \zeta(t) - 2\alpha\zeta(t)$,
and using formula (14.3), we can obtain

$$b_{1\alpha}(t) = \min\left[\exp\left(-\int_0^t \lambda(t')dt'\right)\right]$$

s.t. $r(t) - 2\mu(t) + 2\alpha\mu(t) \leq \lambda(t) \leq s(t) + 2\varepsilon(t) - 2\alpha\varepsilon(t)$,

$$b_{2\alpha}(t) = \max\left[\exp\left(-\int_0^t \lambda(t')dt\right)\right]$$

s.t. $r(t) - 2\mu(t) + 2\alpha\mu(t) \leq \lambda(t) \leq s(t) + 2\varepsilon(t) - 2\alpha\varepsilon(t)$.

Here, as far as the reliability is the monotonically decreasing function, $R(t)$ attains the extreme value at the bounds; we have

(1) when $\alpha \in [0, 0.5)$

$$a_{1\alpha}(t) = \left[\exp\left(-\int_0^t \{s(t') + \varepsilon(t') + \zeta(t') - 2\alpha\zeta(t')\}dt'\right)\right], t > 0,$$
(14.15a)

$$a_{2\alpha}(t) = \left[\exp\left(-\int_0^t \{r(t') - \mu(t') - \nu(t') + 2\alpha\nu(t')\}dt'\right)\right], t > 0,$$
(14.15b)

and (2) $\alpha \in [0.5, 1]$

$$b_{1\alpha}(t) = \left[\exp\left(-\int_0^t \{s(t') + 2\varepsilon(t') - 2\alpha\varepsilon(t')\}dt'\right)\right], t > 0, \quad (14.16a)$$

$$b_{2\alpha}(t) = \left[\exp\left(-\int_0^t \{r(t') - \mu(t') + 2\alpha\nu(t')\}dt'\right)\right], t > 0. \quad (14.16b)$$

By taking the inverse of eqn (14.15) and (14.16), we get the shape functions of the degree of membership. The membership function $\mu_{\tilde{R}(t)}$ of fuzzy reliability is obtained as

$$\mu_{R(t)} = \begin{cases} \frac{\ln(y)+\int_0^t (s(t')+\varepsilon(t')+\zeta(t'))dt'}{\int_0^t (\varepsilon(t')+\zeta(t'))dt'}, & \exp[-\int_0^t \{s(t') + \varepsilon(t') + \zeta(t')\}dt'] \\ \leq y \leq \exp[-\int_0^t s(t')dt'] \\ \frac{\ln(y)+\int_0^t \{s(t')+2\varepsilon(t')\}dt'}{\int_0^t \{2\varepsilon(t')\}dt'}, & \exp[-\int_0^t \{s(t') + 2\varepsilon(t')\}dt'] \\ \leq y \leq \exp[-\int_0^t s(t')dt'] \\ -\frac{\ln(y)+\int_0^t (r(t')-\mu(t')-\nu(t'))dt'}{\int_0^t (\mu(t')+\nu(t'))dt'}, & \exp[-\int_0^t \{r(t')\}dt'] \\ \leq y \leq \exp[-\int_0^t \{r(t') - \mu(t') - \nu(t')\}dt'] \\ -\frac{\ln(y)+\int_0^t (r(t')-2\mu(t'))dt'}{\int_0^t (2\mu(t'))dt'}, & \exp[-\int_0^t \{r(t')\}dt'] \\ \leq y \leq \exp[-\int_0^t \{r(t') - 2\mu(t')\}dt'] \end{cases}$$

(14.17)

Model 1. If the parameter of failure rate $\tilde{H}(t)$ is constant, i.e., $\tilde{H}(t) = H$ then $r(t) = r$, $s(t) = s$, $\mu(t) = \mu$, $\nu(t) = \nu$, $\varepsilon(t) = \varepsilon$, and $\zeta(t) = \zeta$. Now we have

$$\tilde{H}_\alpha(t) = \begin{cases} [r - \mu - \nu + 2\alpha\nu, s + \varepsilon + \zeta - 2\alpha\zeta], \text{for} \alpha \in [0, 0.5) \\ [r - 2\mu + 2\alpha\mu, s + 2\varepsilon - 2\alpha\varepsilon], \text{for} \alpha \in [0.5, 1]. \end{cases}$$

(14.18)

We obtain the membership function

$$\mu_{R(t)} = \begin{cases} \frac{\ln(y)+(s+\varepsilon+\zeta)t}{(\varepsilon+\zeta)t}, & \exp[-(s + \varepsilon + \zeta)t] \leq y \leq \exp[-st] \\ \frac{\ln(y)+(s+2\varepsilon)t}{2\varepsilon t}, & \exp[-(s + 2\varepsilon)t] \leq y \leq \exp[-st] \\ -\frac{\ln(y)+(r-\mu-\nu)t}{(\mu+\nu)t}, & \exp[-rt] \leq y \leq \exp[-(r - \mu - \nu)t] \\ -\frac{\ln(y)+(r-2\mu)t}{2\mu t}, & \exp[-rt] \leq y \leq \exp[-(r - 2\mu)t] \end{cases}$$

(14.19)

Model 2. If the failure rate $\tilde{H}(t)$ is not constant, then the time-dependent hexagonal fuzzy number $\tilde{H}(t)$ depends on the six parameters $r(t)$, $s(t)$, $\mu(t)$, $\nu(t)$, $\varepsilon(t)$, and $\zeta(t)$.

Let us assume that $(t) = \mu = \text{const.}$, $\nu(t) = \text{const.}$, $\varepsilon(t) = \varepsilon = \text{const.}$, $\zeta(t) = \zeta = \text{const.}$, and $r(t) = ce^{kt}$ $s(t) = de^{lt}$ where k and l are positive constants. Since $R(0) = 1$ and $R(\infty) = 0$, from eqn (14.17), we get

$$
\mu_{R(t)} =
\begin{cases}
\frac{\ln(y)+\frac{d}{l}(e^{lt}-1)+(\varepsilon+\zeta)t}{(\varepsilon+\zeta)t}, & \exp[-\frac{d}{l}(e^{lt}-1)-(\varepsilon+\zeta)t] \\
\quad \leq y \leq \exp[-\frac{d}{l}(e^{lt}-1)] \\
\frac{\ln(y)+\frac{d}{l}(e^{lt}-1)+2\varepsilon t}{2\varepsilon t}, & \exp[-\frac{d}{l}(e^{lt}-1)-2\varepsilon t] \\
\quad \leq y \leq \exp[-\frac{d}{l}(e^{lt}-1)] \\
-\frac{\ln(y)+\frac{c}{k}(e^{kt}-1)-(\mu-\nu)t}{(\mu+\nu)t}, & \exp[-\frac{c}{k}(e^{kt}-1)] \\
\quad \leq y \leq \exp[-\frac{c}{k}(e^{kt}-1)+(\mu-\nu)t] \\
-\frac{\ln(y)+\frac{c}{k}(e^{kt}-1)-2\mu t}{2\mu t}, & \exp[-\frac{c}{k}(e^{kt}-1)] \\
\quad \leq y \leq \exp[-\frac{c}{k}(e^{kt}-1)+2\mu t]
\end{cases}
. \quad (14.20)
$$

14.5 Reliability of Different Systems under Fuzzy Conditions where Components have Time-dependent Hexagonal Fuzzy Failure Rate

14.5.1 Series system

The failure rate of the kth component of a system followed by a time-dependent hexagonal fuzzy number is given by a system containing "n" components connected in series.

$\tilde{\lambda}_k(t) = (r_k(t) - \mu_k(t) - \nu_k(t), r_k(t) - \mu_k(t), r_k(t), s_k(t), s_k(t) + \varepsilon_k(t), s_k(t) + \varepsilon_k(t) + \zeta_k(t))$.

Fuzzy reliability of the kth component at time t is $\tilde{R}_k(t)$ for $k = 1, 2, \ldots, n$.

Fuzzy reliability can be expressed as follows in the above n-component series system:

$$
\tilde{R}_S(t) = \prod_{k=1}^{n} \tilde{R}_n(t). \quad (14.21)
$$

From eqn (14.10), we have

$\tilde{R}_k(t) = \exp\left[-\int_0^t \tilde{\lambda}_{ik}(t')dt'\right]$; $k = 1, 2, \ldots, n, t > 0.$

Accordingly, the reliability of the above system is as follows:

$$
\tilde{R}_S(t) = \exp\left[-\int_0^t \left(\sum_{k=1}^{n} \tilde{\lambda}_k(t')dt'\right)\right] = \exp\left[-\int_0^t \tilde{\lambda}_S(t')dt'\right], \quad (14.22)
$$

where $\tilde{\lambda}_S(t) = \sum_{k=1}^{n} \tilde{\lambda}_k(t)$ is the function of the fuzzy failure rate of the series system.

For the membership function, α-cut of $\tilde{\lambda}_k(t)$ is

$$\tilde{\lambda}_k(t, \alpha) = \begin{cases} [r_k(t) - \mu_k(t) - \nu_k(t) + 2\alpha\nu_k(t), \, s_k(t) + \varepsilon_k(t) + \\ \quad \zeta_k(t) - 2\alpha\zeta_k(t)], \, \text{for}\alpha \in [0, 0.5) \\ [r_k(t) - 2\mu_k(t) + 2\alpha\mu_k(t), \, s_k(t) + 2\varepsilon_k(t) \\ \quad -2\alpha\varepsilon_k(t)], \, \text{for}\alpha \in [0.5, 1] \end{cases}.$$

Hence, the α-cut of fuzzy failure rate $\tilde{\lambda}_S(t)$ is given by

$$\tilde{\lambda}_S(t, \alpha) = \begin{cases} [\sum_{k=1}^{n}\{r_k(t) - \mu_k(t) - \nu_k(t) + 2\alpha\nu_k(t)\}, \\ \sum_{k=1}^{n}\{s_k(t) + \varepsilon_k(t) + \zeta_k(t) - 2\alpha\zeta_k(t)\}], \, \text{for}\alpha \in [0, 0.5) \\ [\sum_{k=1}^{n}\{r_k(t) - 2\mu_k(t) + 2\alpha\mu_k(t)\}, \\ \sum_{k=1}^{n}\{s_k(t) + 2\varepsilon_k(t) - 2\alpha\varepsilon_k(t)\}], \, \text{for}\alpha \in [0.5, 1] \end{cases}.$$

Now since $\tilde{R}_S(t)$ is also a hexagonal fuzzy number, we can have the α-cut for the membership function as follows:

(1) When $\alpha \epsilon [0,0.5)$

$$\tilde{R}_S(t, \alpha) = \exp\left[-\int_0^t \left(\sum_{k=1}^{n}\{s_k(t') + \varepsilon_k(t') + \zeta_k(t')\right.\right.$$
$$\left.\left. - 2\alpha\zeta_k(t')\}dt', \right.\right.$$

$$\tilde{R}_S(t, \alpha) = \exp\left[-\int_0^t \left(\sum_{k=1}^{n}\{r_k(t') - \mu_k(t') - \nu_k(t')\right.\right.$$
$$\left.\left. + 2\alpha\nu_k(t')\}dt'. \right.\right.$$

(2) When $\alpha \epsilon [0.5,1]$

$$\tilde{R}_S(t, \alpha) = \exp\left[-\int_0^t \left(\sum_{k=1}^{n}\{s_k(t') + 2\varepsilon_k(t') - 2\alpha\varepsilon_k(t')\}dt'\right)\right],$$

$$\tilde{R}_S(t, \alpha) = \exp\left[-\int_0^t \left(\sum_{k=1}^{n}\{r_k(t') - 2\mu_k(t') + 2\alpha\mu_k(t')\}dt'\right)\right].$$

The failure rate $\tilde{\lambda}_S(t)$ is constant from the model-1. Then the α-cut $\tilde{R}_S(t)$ is given by

$$\tilde{R}_S(t, \alpha) = \left[\begin{array}{c} \exp\{-t\left(\sum_{k=1}^{n}\{s_k + \varepsilon_k + \zeta_k - 2\alpha\zeta_k\}\right)\}, \\ \exp\{-t\left(\sum_{k=1}^{n}\{r_k - \mu_k - \nu_k + 2\alpha\nu_k\}\right)\} \end{array} \right] \text{ for } \alpha \in [0, 0.5)$$

$$(14.23a)$$

$$\tilde{R}_S(t, \alpha) = \left[\begin{array}{c} \exp\{-t\left(\sum_{k=1}^{n}\{s_k + 2\varepsilon_k - 2\alpha\varepsilon_k\}\right)\}, \\ \exp\{-t\left(\sum_{k=1}^{n}\{r_k - 2\mu_k + 2\alpha\mu_k\}\right)\} \end{array} \right] \text{ for } \alpha \in [0.5, 1].$$

(14.23b)

The failure rate $\tilde{\lambda}_S(t)$ is not constant from the model-2; then the α-cut of $\tilde{R}_S(t)$ is obtained as

$$\tilde{R}_S(t, \alpha) = \left[\begin{array}{c} \exp\left\{-\left(\sum_{k=1}^{n}\{\frac{d_k}{l_k}(e^{l_k t} - 1) + t(\varepsilon_k + \zeta_k - 2\alpha\zeta_k)\}\right)\right\}, \\ \exp\left\{-\left(\sum_{k=1}^{n}\{\frac{c_k}{k_k}(e^{k_k t} - 1) - t(\mu_k + \nu_k - 2\alpha\nu_k)\}\right)\right\} \end{array} \right]$$

for $\alpha \in [0, 0.5)$

$$\tilde{R}_S(t, \alpha) = \left[\begin{array}{c} \exp\left\{-\left(\sum_{k=1}^{n}\{\frac{d_k}{l_k}(e^{l_k t} - 1) + 2t(\varepsilon_k - \alpha\varepsilon_k)\}\right)\right\}, \\ \exp\left\{-\left(\sum_{k=1}^{n}\{\frac{c_k}{k_k}(e^{k_k t} - 1) - 2t(\mu_k - \alpha\mu_k)\}\right)\right\} \end{array} \right]$$

for $\alpha \in [0.5, 1].$

14.5.2 Parallel system

Assume that a system with "n" components is connected in parallel and that the failure rate parameter of the kth component of the system is thought of as a time-dependent hexagonal fuzzy number as

$$\tilde{\lambda}_k(t) = (r_k(t) - \mu_k(t) - \nu_k(t), r_k(t) - \mu_k(t), r_k(t), s_k(t), s_k(t) + \varepsilon_k(t),$$

$$q_k(t) + \varepsilon_k(t) + \zeta_k(t)).$$

Let the fuzzy reliability of the kth component at time t is $\tilde{R}_k(t)$ for $k = 1, 2,$ ..., n.

It is well known that the fuzzy reliability of the above-mentioned system $\tilde{R}_P(t)$ is as follows:

$$\tilde{R}_P(t) = 1 - \prod_{k=1}^{n} \left(1 - \tilde{R}_k(t)\right) = 1 - \prod_{k=1}^{n} \left(1 - \exp\left(-\int_0^t \tilde{\lambda}_k(t')dt'\right)\right).$$

Now since $\tilde{R}_P(t)$ is also a hexagonal fuzzy number, we can have the α-cut for fuzzy reliability of parallel system as

(1) When $\alpha \epsilon [0, 0.5)$

$$\tilde{R}_P(t, \alpha) = 1 - \prod_{k=1}^{n} \left(1 - \exp \left\{ - \int_0^t (\{s_k(t') + \varepsilon_k(t') + \zeta_k(t')) \right) \right\}$$
$$- 2\alpha \zeta_k(t')\} dt',$$

$$\tilde{R}_P(t, \alpha) = 1 - \prod_{k=1}^{n} \left(1 - \exp \left\{ - \int_0^t (\{r_k(t') - \mu_k(t') - \nu_k(t')) \right\} \right)$$
$$+ 2\alpha \nu_k(t')\} dt'.$$

(2) When $\alpha \epsilon [0.5, 1]$

$$\tilde{R}_P(t, \alpha) = 1 - \prod_{k=1}^{n} \left(1 - \exp \left\{ - \int_0^t (\{s_k(t') + 2\varepsilon_k(t')) \right\} \right)$$
$$- 2\alpha \varepsilon_k(t')\} dt'$$

$$\tilde{R}_P(t, \alpha) = 1 - \prod_{k=1}^{n} \left(1 - \exp \left\{ - \int_0^t (\{p_k(t') - 2\gamma_k(t')) \right\} \right)$$
$$+ 2\alpha \gamma_k(t')\} dt'.$$

The failure rate is constant in model-1; then the α-cut $\tilde{R}_P(t)$ is given by

$$\tilde{R}_P(t, \alpha) =$$
$$\left[\begin{array}{c} 1 - \prod_{k=1}^{n} (1 - \exp\{-t\{s_k + \varepsilon_k + \zeta_k - 2\alpha\zeta_k\}\}), \\ 1 - \prod_{k=1}^{n} (1 - \exp\{-t\{r_k - \mu_k - \nu_k + 2\alpha\nu_k\}\}) \end{array} \right] \text{ for } \alpha \in [0, 0.5)$$

$$(14.24a)$$

$$\tilde{R}_P(t, \alpha) =$$
$$\left[\begin{array}{c} 1 - \prod_{k=1}^{n} (1 - \exp\{-t\{s_k + 2\varepsilon_k - 2\alpha\varepsilon_k\}\}), \\ 1 - \prod_{k=1}^{n} (1 - \exp\{-t\{r_k - 2\mu_k + 2\alpha\mu_k\}\}) \end{array} \right] \text{ for } \alpha \in [0.5, 1].$$

$$(14.24b)$$

The failure rate function is not constant in model-2; then the α-cut of $\tilde{R}_P(t)$ is obtained as

$$\tilde{R}_P(t,\alpha) =$$

$$\left[\begin{array}{l} 1 - \prod_{k=1}^{n}\left(1 - \exp\left\{-\{\frac{d_k}{l_k}(e^{l_k t} - 1) + t(\varepsilon_k + \zeta_k - 2\alpha\zeta_k)\}\right\}\right), \\ 1 - \prod_{k=1}^{n}\left(1 - \exp\left\{-\{\frac{c_k}{k_k}(e^{k_k t} - 1) - t(\mu_k + \nu_k - 2\alpha\nu_k)\}\right\}\right) \end{array}\right]$$

for $\alpha \in [0, 0.5)$

$$\tilde{R}_P(t,\alpha) =$$

$$\left[\begin{array}{l} 1 - \prod_{k=1}^{n}\left(1 - \exp\left\{-\{\frac{d_k}{l_k}(e^{l_k t} - 1) + 2t(\varepsilon_k - \alpha\varepsilon_k)\}\right\}\right), \\ 1 - \prod_{k=1}^{n}\left(1 - \exp\left\{-\{\frac{c_k}{k_k}(e^{k_k t} - 1) - 2t(\mu_k - \alpha\mu_k)\}\right\}\right) \end{array}\right]$$

for $\alpha \in [0.5, 1]$.

Example: A rocket has three free engines and the functioning of all engines is normal for the successful flying of a rocket. Every engine's failure rates are represented by a hexagonal fuzzy number

$$\tilde{\lambda}_1 = (0.001, 0.0014, 0.0017, 0.002, 0.0024, 0.0027),$$
$$\tilde{\lambda}_2 = (0.0012, 0.0015, 0.0019, 0.0024, 0.0027, 0.003),$$
$$\tilde{\lambda}_3 = (0.0015, 0.0019, 0.0023, 0.0027, 0.003, 0.0033).$$

The system reliability of the rocket using eqn (14.23) is given as follows:

$$\tilde{R}_S(t,\alpha) = \begin{cases} [\exp\{-t(0.009 - 0.0018\alpha)\}, \exp\{-t(0.0037 + 0.0022\alpha)\}] \\ \quad \text{for } \alpha \in [0, 0.5) \\ [\exp\{-t(0.0091 - 0.002\alpha)\}, \exp\{-t(0.0037 + 0.0022\alpha)\}] \\ \quad \text{for } \alpha \in [0.5, 1] \end{cases}$$

The system reliability at time $t = 150$, 200, and 250 is calculated as follows:

$$\tilde{R}_S(150) = (0.25924, 0.29671, 0.34473, 0.412714, 0.486752, 0.57407)$$
$$\tilde{R}_S(200) = (0.1652988, 0.197898, 0.241714, 0.307278, 0.382892, 0.477114)$$
$$\tilde{R}_S(250) = (0.105399, 0.131994, 0.16948, 0.228778, 0.301194, 0.3965314).$$

Example: An airplane has three independent engines and at least one of these works generically for the airplane to fly successfully. Each engine's failure

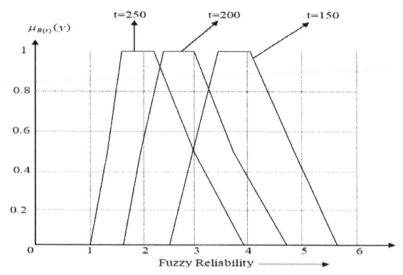

Figure 14.4   Degree of membership of fuzzy reliability for t = 150, 200, and 250.

rates are represented by a hexagonal fuzzy number:

$$\tilde{\lambda}_1 = (0.06, 0.1, 0.15, 0.2, 0.24, 0.29),$$
$$\tilde{\lambda}_2 = (0.05, 0.09, 0.13, 0.17, 0.22, 0.27),$$
$$\tilde{\lambda}_3 = (0.07, 0.11, 0.16, 0.21, 0.25, 0.29).$$

The system reliability of the rocket using eqn (14.24) is given as follows:

$$\tilde{R}_P(t, \alpha) = \left[\begin{array}{c} 1 - (1 - \exp\{-t\{0.29 - 0.1\alpha\}\}) \times \\ (1 - \exp\{-t\{0.27 - 0.1\alpha\}\}) \times \\ (1 - \exp\{-t\{0.29 - 0.08\alpha\}\}), \\ 1 - (1 - \exp\{-t\{0.05 + 0.1\alpha\}\}) \times \\ (1 - \exp\{-t\{0.05 + 0.08\alpha\}\}) \times \\ (1 - \exp\{-t\{0.07 + 0.08\alpha\}\}) \end{array} \right] \quad \text{for } \alpha \in [0, 0.5)$$

$$\tilde{R}_P(t, \alpha) = \left[\begin{array}{c} 1 - (1 - \exp\{-t\{0.28 - 0.08\alpha\}\}) \times \\ (1 - \exp\{-t\{0.27 - 0.1\alpha\}\}) \\ \times (1 - \exp\{-t\{0.29 - 0.08\alpha\}\}), \\ 1 - (1 - \exp\{-t\{0.05 + 0.1\alpha\}\}) \times \\ (1 - \exp\{-t\{0.05 + 0.08\alpha\}\}) \\ \times (1 - \exp\{-t\{0.06 + 0.1\alpha\}\}) \end{array} \right] \quad \text{for } \alpha \in [0.5, 1].$$

The system reliability at times $t = 10$ and 20 is calculated as follows:

$$\tilde{R}_S(10) = (0.167032139, 0.25783752, 0.379835817,$$
$$0.548953326, 0.749746923, 0.922062283)$$
$$\tilde{R}_S(20) = (0.010535218, 0.0270065, 0.065307294,$$
$$0.156218735, 0.358234095, 0.698957927).$$

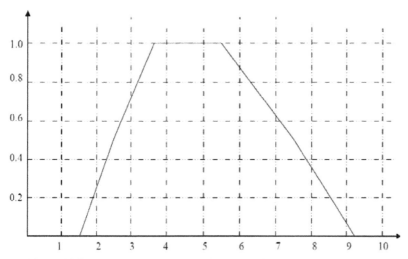

Figure 14.5 Degree of membership function of fuzzy reliability for $t = 10$.

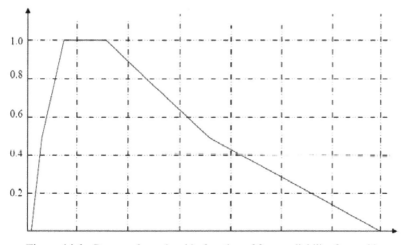

Figure 14.6 Degree of membership function of fuzzy reliability for $t = 20$.

14.6 Conclusion

In this chapter, we defined time-dependent hexagonal fuzzy numbers and α-cut of hexagonal fuzzy numbers. We obtain the reliability function using the extension principle and time-dependent hexagonal fuzzy numbers with two models. The failure rate is constant and the failure rate is not constant in model-1 and model-2, respectively. In fuzzy reliability functions, the degree of membership and non-membership is determined using time-dependent hexagonal fuzzy numbers. Numerical examples are also taken for evaluate fuzzy reliability when the failure rate is expressed as time-dependent hexagonal fuzzy numbers. The fuzzy reliability of series and parallel systems is estimated at various times $t = 150$, 200, and 250, and time $t = 10$ and 20, respectively (shown in Figures 14.4–14.6).

Acknowledgement

The authors sincerely thank and appreciate the editors and reviewers for their time and valuable comments.

References

[1] T. Onisawa and J. Kacprzyk, 'Reliability and safety analysis under fuzziness', Physica-Verlag Heidelberg 1995.

[2] L. A. Zadeh, 'Fuzzy sets', Inform. Control, 8(3), pp.338-353, 1965.

[3] K. Y. Cai, C. Y. Wen and M. L. Zhang, 'Fuzzy variables as a basis for a theory of fuzzy reliability in the possibility context', Fuzzy Sets and Systems, 42, pp.142–145, 1991.

[4] K. Y. Cai, C. Y. Wen, M. L. Zhang, 'Fuzzy states as for a theory of fuzzy reliability', Microelectron Reliability, 33, pp.2253-2263, 1993.

[5] K. Y. Cai, 'System failure engineering and fuzzy methodology. An introductory overview', Fuzzy Set and Systems, 83, pp.113-133, 1996.

[6] M. Aliev and Z. Kara, 'Fuzzy system reliability analysis using time dependent fuzzy set', Control and Cybernetics, 33, pp.653–662, 2004.

[7] M. Kumar, S. P. Yadav and S. Kumar, 'Fuzzy system reliability evaluation using time-dependent intuitionistic fuzzy set', International journal of systems science, 44(1), pp.50-66, 2011.

[8] D. Kumar and A. Kumari, 'An Evaluation of the System Reliability Using Fuzzy Lifetime Distribution Emphasising Hexagonal Fuzzy Number', IOSR Journal of Engineering, 9(8), pp.17-25, 2019.

[9] P. Kumar and S. B. Singh, 'Fuzzy system reliability using intuitionistic fuzzy Weibull lifetime distribution', International Journal of Reliability and Applications, 16(1), pp.15–26, 2015.

[10] S. M. Chen, 'Fuzzy system reliability analysis using fuzzy number arithmetic operations', Fuzzy Sets and Systems, 64, pp.31-38, 1994.

[11] A. S. Sudha, P. Rajarajeshwari and R. Karthika, 'A new Operation on Hexagonal Fuzzy Number', International Journal of Fuzzy Logic Systems, 3(3), pp.15-26, 2013.

[12] K. Dhurai and A. Karpagam, 'A new membership function on hexagonal fuzzy numbers', International journal of science and research, 5(6), pp.1129-1131, 2016.

[13] G. S. Mahapatra, T. K. Roy, 'Reliability evaluation using triangular intuitionistic fuzzy numbers arithmetic operations', World Academy of Science and Technology, Vol. 50, pp. 574-581, 2009.

[14] A. K. Shaw and T. K. Roy, 'Trapezoidal Intuitionistic Fuzzy Number with some arithmetic operations and its application on reliability evaluation', International Journal Mathematics in Operational Research, 5, 55-73, (1993).

Index

About the Editors

Shristi Kharola is currently working toward the Ph.D. degree in mathematics from Graphic Era Deemed to be University, Dehradun, India. Prior to arriving at Graphic Era University, Shristi completed a three-year graduate honors degree in mathematics from Graphic Era Hill University, Dehradun, India, and postgraduation degree in mathematics from H. N. B. Garhwal Central University, India. Her research interests include areas of reliability engineering, fuzzy logic, decision-making, applied-mathematics, and circular economy, ranging from theory to design to implementation. In the years 2020−2021, Shristi collaborated actively with countries, Turkey and UK, for an international project jointly with Government of India. Shristi has published some good articles with Elsevier, Springer, etc., and has presented her publications in seminars, workshops, and conferences globally.

Mangey Ram received the Ph.D. degree major in mathematics and minor in computer science from the G. B. Pant University of Agriculture and Technology, Pantnagar, India, in 2008. He has been a Faculty Member for around 13 years and has taught several core courses in pure and applied mathematics at undergraduate, postgraduate, and doctorate levels. He is currently the Research Professor with Graphic Era Deemed to be University, Dehradun, India, and the Visiting Professor with Peter the Great St. Petersburg Polytechnic University, Saint Petersburg, Russia. Before joining the Graphic Era, he was a Deputy Manager (Probationary Officer) with Syndicate Bank for a short period. He is the Editor-in-Chief for *International Journal of Mathematical, Engineering and Management Sciences*, *Journal of Reliability and Statistical Studies*, and *Journal of Graphic Era University*; Series Editor of six book series with Elsevier, CRC Press-A Taylor and Francis Group, Walter De Gruyter Publisher Germany, and River Publisher, and the Guest Editor and Associate Editor for various journals. He has published more than 300 publications (journal articles/books/book chapters/conference articles) in IEEE, Taylor & Francis, Springer Nature, Elsevier, Emerald, and World Scientific and many other national and international journals and conferences. Also, he has published more than 60 books (authored/edited) with international publishers like Elsevier, Springer Nature, CRC Press-A Taylor and Francis Group, Walter De Gruyter Publisher Germany, and River Publisher. His fields of research are reliability theory and applied mathematics. Dr. Ram is a Senior Member of the IEEE, Senior Life Member of Operational Research Society of India, Society for Reliability Engineering, Quality and Operations Management in India, and Indian Society of Industrial and Applied Mathematics, He has been a member of the organizing committee of a number of international and national conferences, seminars, and workshops. He has been conferred with "Young Scientist Award" by the Uttarakhand State Council for Science and Technology, Dehradun, India, in 2009. He has been awarded the "Best Faculty Award" in 2011; "Research Excellence Award" in 2015; and "Outstanding Researcher Award" in 2018 for his significant contribution in academics and research at Graphic Era Deemed to be University, Dehradun, India. Recently, he has received the "Excellence in Research of the Year-2021 Award" by the Honorable Chief Minister of Uttarakhand State, India.

Sachin K. Mangla is working in the field of green and sustainable supply chain and operations; Industry 4.0; circular economy; decision making and simulation. He is committed to do and promote high quality research. He has published/presented several papers in reputed international/national journals (*Journal of Business Research*; *International Journal of Production Economics*; *International Journal of Production Research*; *Computers and Operations Research*; *Production Planning and Control*; *Business Strategy and the Environment*; *Annals of Operations Research*; *Transportation Research Part-D*; *Transportation Research Part-E*; *Renewable and Sustainable Energy Reviews*; *Resource Conservation and Recycling*; *Information System Frontier*; *Journal of Cleaner Production*; *Management Decision*; *Industrial Data and Management System*) and conferences (POMS, SOMS, IIIE, CILT–LRN, MCDM, and GLOGIFT). He has an h-index 55, i10-index 96, Google Scholar Citations of more than 8000. He is involved in several editorial positions and editing couple of special issues as a Guest Editor in top tier journals. Currently, he is working as an Associate Editor for the *Journal of Cleaner Production*, *International Journal of Logistics Management*, *Sustainable Production and Consumption*, and *IMA Journal of Management Mathematics*. He is also involved in several research projects on various issues and applications of circular economy, Industry 4.0, and sustainability. Among them, he also contributed to the knowledge-based decision model in "Enhancing and implementing knowledge-based ICT solutions within high risk and uncertain conditions for agriculture production systems (RUC-APS)," European Commission RISE scheme, âĆň1.3M. Recently, he has also received a grant as a PI from British Council – Newton Fund Research Environment Links Turkey/UK – Circular and Industry 4.0 driven solutions for reducing food waste in supply chains. He is also working in several projects on food waste and circular economy and Industry 4.0 issues, sponsored by government of India (ICSSR) and state government agencies (USERC).

Yigit Kazancoglu received the B.S. degree from Eastern Mediterranean University, Department of Industrial Engineering. He has graduated from Coventry University, MBA and Izmir University of Economics MBA programs. He received the Ph.D. degree from Ege University (Production & Operations Management). He is a full Professor Doctor with Yasar University, Logistics Management Department and Head of the department. His research areas are in operations management and supply chain management. The subjects he is working on are sustainability, circular economy, green supply chain management, and digitalization. He has published more than 100 articles in the SCI, SCI-E, SSCI, ESCI, and SCOPUS indexed journals. Prof Dr. Kazancoglu has papers on related topics that were published in *International Journal of Production Research*, *Production Planning and Control*, *Business Strategy and the Environment*, *Journal of Cleaner Production*, *International Journal of Logistics Management*, *Expert Systems with Applications*, *Annals of Operations Research*, *Transportation Research Part – E*, *Technological Forecasting and Social Change*, *Sustainable Cities and Society*, *Sustainable Production and Consumption*, *Resource Conservation and Recycling*, *Environmental Science and Pollution Research*, *Sustainable Development*, *Journal of Material Cycles and Waste Management*, *Computers and Operations Research*, *Journal of Global Information Management*, *Science of Total Environment*, *Operations Management Research*, *Journal of Enterprise Information Management*, *International Journal of Logistics Research and Applications*, *Industrial Data and Management System*, and *International Journal of Quality and Reliability Management*. He is an Associate Editor for *International Journal of Mathematical, Engineering and Management Sciences* (IJMEMS) (Scopus, ESCI), special issue Guest Editor for *Journal of Enterprise Information Management* (JEIM), "Data driven design and management of sustainability for emerging modes of supply chain," *International Journal of Productivity and Performance Management*

(IJPPM) "Circular economy and sustainable business performance management in the era of digitalization," *Journal of Business Research* "AI and data analytics for omnichannel health care business," *Sustainability* "Smart and sustainable food supply chain management," *International Journal of Global Business and Competitiveness* "The competitiveness of SMEs in the era of circular and digital economy," and *Computers & Industrial Engineering* "Carbon neutrality through Industry 4.0 based smart manufacturing." Prof. Dr. Kazancoglu completed eight projects funded by EU under Erasmus+, Transversal, EU, Grundtvig, as well as Newton Fund Research Environment Links Project titled as "Developing capacity and research network on circular and Industry 4.0 driven sustainable solutions for reducing food waste in supply chains in Turkey" and USERC – Indian Govt. Funded Research Project – "Food Waste Management in Circular Economy."